D0025891

Astrophysics Simulations
The Consortium for
Upper-Level Physics Software

J. M. Anthony Danby
Department of Mathematics, North Carolina State University
Raleigh, North Carolina

Richard Kouzes
Pacific Northwest Laboratory
Richland, Washington

Charles Whitney
Harvard-Smithsonian Center for Astrophysics
Cambridge, Massachusetts

Series Editors

Robert Ehrlich

William MacDonald

Maria Dworzecka

JOHN WILEY & SONS, INC.
NEW YORK · CHICHESTER · BRISBANE · TORONTO · SINGAPORE

ACQUISITIONS EDITOR Cliff Mills
MARKETING MANAGER Catherine Faduska
PRODUCTION EDITOR Sandra Russell
DESIGNER Maddy Lesure
MANUFACTURING MANAGER Susan Stetzer

This book was set in 10/12 Times Roman by Beacon Graphics and
printed and bound by Hamilton Printing. The cover was printed by Phoenix Color.

Recognizing the importance of preserving what has been written, it is a
policy of John Wiley & Sons, Inc. to have books of enduring value published
in the United States printed on acid-free paper, and we exert our best
efforts to that end.

The paper on this book was manufactured by a mill whose forest management programs include
sustained yield harvesting of its timberlands. Sustained yield harvesting principles ensure that
the number of trees cut each year does not exceed the amount of new growth.

Copyright © 1995, by John Wiley & Sons, Inc.

All rights reserved. Published simultaneously in Canada.

Reproduction or translation of any part of
this work beyond that permitted by Sections
107 and 108 of the 1976 United States Copyright
Act without the permission of the copyright
owner is unlawful. Requests for permission
or further information should be addressed to
the Permissions Department, John Wiley & Sons, Inc.

Library of Congress Cataloging in Publication Data:
Danby, J. M. A.
    Astrophysics simulations : the Consortium for Upper Level Physics
Software / J. M. Anthony Danby, Richard Kouzes, Charles Whitney.
       p.   cm.
    Includes bibliographical references (p.).
    ISBN 0-471-54879-0 (pbk./disk)
    1. Astronomy—Data processing.   2. Astronomy—Computer simulation.
I. Kouzes, Richard.   II. Whitney, Charles.   III. Consortium for
Upper Level Physics Software.   IV. Title.
QB51.3.E43D26   1995
523.01'01'13—dc20   94-38485
                                                              CIP

Printed in the United States of America

10 9 8 7 6 5 4 3 2 1

This remarkable relationship, whereby mass alone determines the equilibrium of a star, is the reason that the color-brightness diagram is generally considered the most convenient map for discussing stellar physics and comparing stars of different types.

But how does the star of a particular mass "decide" what color and brightness to have? This is the question answered by the stellar interior program (chapter 5), which constructs models for stars on this zero-age line.

The simulation integrates the differential equations governing stellar structure to reach an equilibrium condition, thereby determining the details of the interior of the star. The user can explore the effect on the equilibrium produced by the various stellar parameters.

Stellar Evolution

As the star ages, its radius and luminosity change and the star traces out a trajectory on the color-brightness diagram.

The evolution program (chapter 6) shows how this takes place. The calculations are highly simplified, but they carry a young star along the so-called Hyashi track to the zero-age main sequence, and they carry a star off the main sequence to the stage of shell burning. It is still impractical to carry out a complete set of calculations on a desktop computer, so the remaining stages are depicted with animations.

The section on getting started with the program and the suggested exercises lead the user through the initial usage of the programs.

Temporal Brightness Variations

In addition to the explosive brightening of novae, some stars show smooth and regular variations of brightness. These regular variations are caused by bodily pulsation (spherical expansion and contraction), and the resulting curves of variation are nearly sinusoidal. These are called the pulsating stars, or Cepheid variables.

Chapter 7 describes the simulation of a pulsating star. This simulation uses the equations of motion of spherical elastic shells to generate the possible motions of a model star, and it displays them in a variety of graphs. By adjusting the mass, radius, and luminosity of the model, the user may display the pulsation characteristics of a wide range of stars from the Sun to a supergiant.

Stellar Atmospheres

Hot gases deep inside a star produce x-ray photons that diffuse toward the surface—some photons are deflected by electron scattering; others are absorbed by atoms. When an atom absorbs a photon, the energy is usually re-emitted as a collection of less energetic photons, and the outward-diffusing x-rays are gradually converted to ultraviolet light. By the time they have arrived at the surface, the photons typically consist of visible light. The pattern of colors in the spectrum of photons that escape from the surface is, for some stars such as the Sun, a good imitation of the blackbody curve described by Planck's law of radiation. The overall shape of the spectrum is a clue to the surface temperature of the star.

Each species of gas in the atmosphere leaves its earmarks in the emitted spectrum, and the interpretation of these features provides clues to the chemical composition and to the pressure of the atmosphere. These, in turn, provide clues to the mass, radius, and age of the star. So the spectrum, when interpreted in terms of a model atmosphere, can reveal a great deal about the history of individual stars.

The simulation described in chapter 8 permits displaying the continuous spectrum of two stars at once, thereby comparing them. It also permits superimposing a blackbody curve. The color-brightness diagrams of selected constellations may also be displayed.

These simulations should be used independently at the point where they fit into a course sequence. Each chapter gives an introduction to the concepts related to the simulation, followed by a description of the mathematics used, and instructions on using the program. Users are encouraged to read the chapter introductions and gain some experience running the programs before trying to understand the detailed mathematics.

Preface

During the past several decades, the applications of physics to astronomy have mushroomed—even cosmology has entered the ranks of physical science and is no longer viewed as a purely philosophical topic. Hence, astrophysics has come to include a tremendous variety of subjects and techniques. The selection of topics presented here, though very limited, is, we feel, appropriate to an undergraduate course; and although we do not apologize for this narrowness, we would like to explain briefly what we have attempted to achieve. We would also like to add a few words on the manner in which this book and associated software might be used.

A typical course in astrophysics is highly eclectic; it is designed and given by someone who has a favorite set of topics. There appears to be no standard course in introductory astrophysics, nor have we set out to create one. We have followed this convention in that we have treated topics that are of special interest to us but are also representative of the theoretical techniques of modern astrophysics. We have also tried to choose topics that lend themselves to computer simulations.

This book does not represent our ideal textbook. Although it could be used as a basis for a course in stellar astrophysics, we prefer to think of it as a resource book, to be used in conjunction with other materials assembled by the instructor.

The simulations in this course are related to two major areas in astronomy and astrophysics:

1. dynamics, relating to binary stars, the motion of *n*-bodies, and galactic rotation; and

2. stellar physics, that is, building models of various types of stars and their atmospheres.

The simulations may be used in rather different ways by teachers and by students. Teachers may use them as lecture demonstrations in which the behaviors of various systems are illustrated and described qualitatively. Used this way, they are appropriate for introductory courses and they do not require much in the way of mathematics or physics on the part of the students. These simulations could provide enrichment for a conventional physics course. They will also provide evidence for the importance of modeling in astrophysics and they will indicate the variey of models that are used in stellar physics.

On the other hand, upper-level physics students who have had conventional courses in mechanics, thermodynamics, and ordinary and partial differential equations can use these simulations for exploration and for homework assignments. The code is written in Pascal, so it is fairly easy to read and modify. Students can either use the programs as they stand, or they may change some of the key subroutines and see the consequences.

The following is a brief description of the topics treated by these simulations. For more details, see the individual chapters.

Part 1: Dynamics

Virtually all our quantitative knowledge of stellar parameters, such as mass and true brightness, come from studies of multiple systems of stars. In addition, these systems provide examples of mass transfer and the formation of accretion disks. The simulations are intended to provide demonstrations of the principal phenomena associated with the three types of binaries. There is also a program illustrating phenomena in the restricted problem of three bodies, of which some knowledge is essential in order to understand mass transfer. Another simulation follows the possible orbital evolution of a close binary system under the influence of tides.

Many astrophysical phenomena can only be investigated by computing the motion of systems of attracting masses. One aim of the simulations is to bring this material into the classroom and the elementary curriculum. Two basic models are used. In the first, a system of *n*-bodies moves subjects to their Newtonian attraction. In the second, a system of "massless" bodies moves in the gravitational field of a binary system. In each case, two different models have been chosen.

The simulations of galactic kinematics illustrate the rotation of a galaxy, with the use of the 21-cm line to find galactic structure, and the local motion of stars close to the Sun. Users have considerable latitude in the design of the galaxy.

Much of the material in the text is included so that a user can understand the construction of the program. It is our hope that users will modify the code freely, and for this, a basic understanding of the theory is essential. Detailed knowledge of the theory is not essential in order to profit from the simulations.

Part 2: Young Stars on the Color-Magnitude Diagram

Stellar Structure

Some stars are much brighter than others; some are redder and some bluer. If you make a plot in which the true brightness of a star is taken as the *y*-axis and its color is taken as the *x*-axis (conventionally blue to the left, red to the right) you will produce what is called a "color-brightness diagram." The stars in the neighborhood of the Sun lie on a single line on this diagram, known as the main sequence. From this simple fact, it has been concluded that a single parameter determines where a young star lies on the diagram—this crucial parameter is the mass of the star. If a blob of matter of a certain mass settles down into a star, it will lie at a particular point along this line. Another blob with a higher mass will lie higher and farther to the left on the diagram.

Contents

List of Figures

1

Introduction

"It is nice to know that the computer understands the problem. But I would like to understand it too."

—Eugene P. Wigner, quoted in _Physics Today,_ July 1993

1.1   Using the Book and Software

The simulations in this book aim to exploit the capabilities of personal computers and provide instructors and students with valuable new opportunities to teach and learn physics, and help develop that all-important, if somewhat elusive, physical intuition. This book and the accompanying diskettes are intended to be used as supplementary materials for a junior- or senior-level course. Although you may find that you can run the programs without reading the text, the book is helpful for understanding the underlying physics, and provides numerous suggestions on ways to use the programs. _If you want a quick guided tour through the programs, consult the "Walk Throughs" in Appendix A._ The individual chapters and computer programs cover mainstream topics found in most textbooks. However, because the book is intended to be a supplementary text, no attempt has been made to cover all the topics one might encounter in a primary text.

Because of the book's organization, students or instructors may wish to deal with different chapters as they come up in the course, rather than reading the chapters in the order presented. One price of making the chapters semi-independent of one another is that they may not be entirely consistent in notation or tightly cross-referenced. Use of the book may vary according to the taste of the student or instructor. Students may use this material as the basis of a self-study course. Some instructors may make homework assignments from the large number of exercises in each chapter or to use them as the basis of student projects. Other instructors may use the computer programs primarily for in-class demonstrations. In this latter case, you may find that the programs are suitable for a range of courses from the introductory to the graduate level.

Use of the book and software may also vary with the degree of computer programming performed by users. For those without programming experience, all the computer simulations have been supplied in executable form, permitting them to be used as is. On the other hand, Pascal source code for the programs has also been provided, and a number of exercises suggest specific ways the programs can be modified. Possible modifications range from altering a single procedure especially set up for this purpose by the author, to larger modifications following given examples, to extensive additions for ambitious projects. However, the intent of the authors is that the simulations will help the student to develop intuition and a deeper understanding of the physics, rather than to develop computational skills.

We use the term "simulations" to refer to the computer programs described in the book. This term is meant to imply that programs include complex, often realistic, calculations of models of various physical systems, and the output is usually presented in the form of graphical (often animated) displays. Many of the simulations can produce numerical output—sometimes in the form of output files that could be analyzed by other programs. The user generally may vary many parameters of the system, and interact with it in other ways, so as to study its behavior in real time. The use of the term simulation should not convey the idea that the programs are bypassing the necessary physics calculations and simply producing images that look more or less like the real thing.

The programs accompanying this book can be used in a way that complements, rather than displaces, the analytical work in the course. It is our belief that, in general, computational and analytical approaches to physics can be mutually reinforcing. It may require considerable analytical work, for example, to modify the programs, or really to understand the results of a simulation. In fact, one important use of the simulations is to suggest conjectures that may then be verified, modified, or proven false analytically. A complete list of programs is given in Section 1.7.

1.2   Required Hardware and Installation of Programs

The programs described in this book have been written in the Pascal language for MS-DOS platforms. The language is Borland/Turbo Pascal, and the minimum hardware configuration is an IBM-compatible 386-level machine preferably with math coprocessor, mouse, and VGA color monitor. In order to accommodate a wide range of machine speeds, most programs that use animation include the capability to slow down or speed up the program. To install the programs, place disk number 1 in a floppy drive. Change to that drive, and type Install. You need only type in the file name to execute the program. Alternatively, you could type the name of the driver program (the same name as the directory in which the programs reside), and select programs from a menu. A number of programs write to temporary files, so you should check to see if your autoexec.bat file has a line that sets a temporary directory, such as SET TEMP = C:\TEMP. (If you have installed WINDOWS on your PC, you will find that such a command has already been written into your autoexec.bat file.) If no such line is there, you should add one.

Compilation of Programs

If you need to compile the programs, it would be preferable to do so using the Borland 7.0 (or later) compiler. If you use an earlier Turbo compiler you may run out of memory when compiling. If that happens, try compiling after turning off memory resident programs. If your machine has one, be sure to compile with the math-coprocessor turned on (no emulation). Finally, if you recompile programs using any compiler other than Borland 7.0, you will get the message: "EGA/VGA Invalid Driver File" when you try to execute them, because the driver file supplied was produced using this version of the compiler. In this case, search for the file BGILINK.pas included as part of the compiler to find information on how to create the EGAVGA.obj driver file. *If any other instructions are needed for installation, compilation, or running of the programs, they will be given in a README file on the diskettes.*

1.3   User Interface

To start a program, simply type the name of the individual or driver program, and an opening screen will appear. All the programs in this book have a common user interface. Both keyboard and mouse interactions with the computer are possible. Here are some conventions common to all the programs.

Menus: If using the *keyboard*, press **F10** to highlight one of menu boxes, then use the **arrow** keys, **Home**, and **End** to move around. When you press **Return** a submenu will pull down from the currently highlighted menu option. Use the same keys to move around in the submenu, and press **Return** to choose the highlighted submenu entry. Press **Esc** if you want to leave the menu without making any choices.

   If using the *mouse* to access the top menu, click on the menu bar to pull down a submenu, and then on the option you want to choose. Click anywhere outside the menus if you want to leave them without making any choice. Throughout this book, the process of choosing submenu entry **Sub** under main menu entry **Main** is referred to by the phrase "choose **Main | Sub**." The detailed structure of the menu will vary from program to program, but all will contain **File** as the first (left-most) entry, and under **File** you will find **About CUPS, About Program, Configuration,** and **Exit Program**. The first two items when activated by mouse or arrows keys will produce information screens. Selecting **Exit Program** will cause the program to terminate, and choosing **Configuration** will present you with a list of choices (described later), concerning the mode of running the program. In addition to these four items under the **File** menu, some programs may have additional items, such as **Open**, used to open a file for input, and **Save**, used to save an output file. If **Open** is present and is chosen, you will be presented with a scrollable list of files in the current directory from which to choose.

Hot Keys: Hot keys, usually listed on a bar at the bottom of the screen, can be activated by pressing the indicated key or by clicking on the hot key bar with the mouse. The hot key **F1** is reserved for help, the hot key **F10** activates the menu bar. Other hot keys may be available.

Sliders (scroll-bars): If using the *keyboard*, press **arrow** keys for slow scrolling of the slider, **PgUp/PgDn** for fast scrolling, and **End/Home** for moving from one end to another. If you have more then one slider on the screen then only the slider with marked "thumb" (sliding part) will respond to the above keys. You can toggle the mark between your sliders by pressing the **Tab** key.

If using the *mouse* to adjust a slider, click on the thumb of the slider, drag it to desired value, and release. Click on the arrow on either end of the slider for slow scrolling, or in the area on either side of thumb for fast scrolling in this direction. Also, you can click on the box where the value of the slider is displayed, and simply type in the desired number.

Input Screens: All input screens have a set of "default" values entered for all parameters, so that you can, if you wish, run the program by using these original values. Input screens may include circular radio buttons and square check boxes, both of which can take on Boolean, i.e., "on" or "off," values. Normally, check boxes are used when only one can be chosen, and radio buttons when any number can be chosen.

If using the *keyboard*, press **Return** to accept the screen, or **Esc** to cancel it and lose the changes you may have made. To make changes on the input screen by keyboard, use **arrow** keys, **PgUp**, **PgDn**, **End**, **Home**, **Tab**, and **Shift-Tab** to choose the field you want to change, and use the backspace or delete keys to delete numbers. For Boolean fields, i.e., those that may assume one of two values, use any key except those listed above to change its value to the opposite value.

If you use the *mouse*, click [OK] to accept the screen or [Cancel] to cancel the screen and lose the changes. Use the mouse to choose the field you want to change. Clicking on the Boolean field automatically changes its value to the opposite value.

Parser: Many programs allow the user to enter expressions of one or more variables that are evaluated by the program. The function parser can recognize the following functions: absolute value (abs), exponential (exp), integer or fractional part of a real number (int or frac), real or imaginary part of a complex number (re or im), square or square root of a number (sqr or sqrt), logarithms—base 10 or e (log or ln)—unit step function (h), and the sign of a real number (sgn). It can also recognize the following trigonometric functions: sin, cos, tan, cot, sec, csc, and the inverse functions arcsin, arccos, arctan, as well as the hyperbolic functions denoted by adding an "h" at the end of all the preceding functions. In addition, the parser can recognize the constants *pi*, *e*, $i(\sqrt{-1}\,)$, and rand (a random number between 0 and 1). The operations **+, −, *, /,** ^(exponentiation), and !(factorial) can all be used, and the variables *r* and *c* are interpreted as $r = \sqrt{x^2 + y^2}$ and $c = x + iy$. Expressions involving these functions, variables, and constants can be nested to an arbitrary level using parentheses and brackets. For example, suppose you entered the following expression: **h(abs(sin(10*pi* x))−0.5)**. The parser would interpret this function as $h(|sin(10\pi x)|-0.5)$. If the program evaluates this function for a range of *x*-values, the result, in this case, would be a series of square pulses of width 1/15, and center-to-center separation 1/10.

Help: Most programs have context-sensitive help available by pressing the **F1** hot
key (or clicking the mouse in the **F1** hot key bar). In some programs help is
also available by choosing appropriate items on the menu, and in still other
programs tutorials on various aspects of the program are available.

1.4   The CUPS Project and CUPS Utilities

The authors of this book have developed their programs and text as part of the
Consortium for Upper-Level Physics Software (CUPS). Under the direction of the
three editors of this book, CUPS is developing computer simulations and associated texts for nine junior- or senior-level courses, which comprise most of the undergraduate physics major curriculum during those two years. A list of the nine
CUPS courses, and the authors associated with each course, follows this section.
This international group of 27 physicists includes individuals with extensive back-
grounds in research, teaching, and development of instructional software.

   The fact that each chapter of the book has been written by a different author
means that the chapters will reflect that individual's style and philosophy. Every
attempt has been made by the editors to enhance the similarity of chapters, and
to provide a similar user interface in each of the associated computer simulations. Consequently, you will find that the programs described in this and other
CUPS books have a common look and feel. This degree of similarity was made
possible by producing the software in a large group that shared a common philosophy and commitment to excellence.

   Another crucial factor in developing a degree of similarity between all CUPS
programs is the use of a common set of utilities. These CUPS utilities were written
by Jaroslaw Tuszynski and William MacDonald, the former having responsibility
for the graphics units, and the latter for the numerical procedures and functions.
The numerical algorithms are of high quality and precision, as required for reliable results. CUPS utilities were originally based on the M.U.P.P.E.T. utilities of
Jack Wilson and E.F. Redish, which provided a framework for a much expanded
and enhanced mathematical and graphics library. The CUPS utilities (whose
source code is included with the simulations with this book), include additional
object-oriented programs for a complete graphical user interface, including pull-
down menus, sliders, buttons, hotkeys, and mouse clicking and dragging. They
also include routines for creating contour, two-dimensional (2-D) and 3-D plots,
and a function parser. The CUPS utilities have been provided in source code form
to enable users to run the simulations under future generations of Borland/Turbo
Pascal. If you do run under future generations of Turbo or Borland Pascal on the
PC, the utilities and programs will need to be recompiled. You will also need
to create a new egavga.obj file which gets combined with the programs when an
executable version is created—thereby avoiding the need to have separate
(egavga.bgi) driver files. These CUPS utilities are also available to users who
wish to use them for their own projects.

   One element not included in the utilities is a procedure for creating hard copy
based on screen images. When hard copy is desired, those PC users with the appropriate graphics driver (graphics.com), may be able to produce high-quality
screen images by depressing the **PrintScreen** key. If you do not have the graph-
ics software installed to get screen dumps, select **Configuration | Print Screen**,

and follow the directions. Moreover, public domain software also exists for capturing screen images, and for producing PostScript files, but the user should be aware that such files are often quite large, sometimes over 1 MB, and they require a PostScript printer driver to produce.

One feature of the CUPS utilities that can improve the quality of hard copy produced from screen captures is a procedure for switching colors. This capability is important because the gray scale rendering of colors on black-and-white printers may create poor contrasts if the original (default) color assignments are used. To access the CUPS utility for changing colors, the user need only choose **Configuration** under the **File** menu when the program is first initiated, or at any later time. Once you have chosen **Configuration**, to change colors you need to click the mouse on the **Change Colors** bar, and you will be presented with a 16 by 16 matrix of radio buttons that will allow you to change any color to any other color, or else to use predefined color switches, such as a color "reversal," or a conversion of all light colors to black, and all dark colors to white. (The screen captures given in this book were produced using the "reverse" color map.) Any such color changes must be redone when the program is restarted.

Other system parameters may likewise be set from the **File | Configuration** menu item. These include the path for temporary files that the program may create (or want to read), the mouse "double click" speed—important for those with slow reflexes—an added time delay to slow down programs on computers that are too fast, and a "check memory" option—primarily of interest to those making program modifications.

Those users wishing more information on the CUPS utilities should consult the CUPS Utilities Manual, written by Jaroslaw Tuszynski and William MacDonald, published by John Wiley and Sons. However, it is not necessary for casual users of CUPS programs to become familiar with the utilities. Such familiarity would only be important to someone wishing to write their own simulations using the utilities. The utilities are freely available for this purpose, for unrestricted noncommercial production and distribution of programs. However, users of the utilities who wish to write programs for commercial distribution should contact John Wiley and Sons.

1.5   Communicating With the Authors

Users of these programs should not expect that run-time errors will never occur! In most cases, such run-time errors may require only that the user restart the program; but in other cases, it may be necessary to reboot the computer, or even turn it off and on. The causes of such run-time errors are highly varied. In some cases, the program may be telling you something important about the physics or the numerical method. For example, you may be trying to use a numerical method beyond its range of applicability. Other types of run-time errors may have to do with memory or other limitations of your computer. Finally, although the programs in this book have been extensively tested, we cannot rule out the possibility that they may contain errors. (Please let us know if you find any! It would be most helpful if such problems were communicated by electronic mail, and with complete specificity as to the circumstances under which they arise.)

It would be best if you communicated such problems directly to the author of each program, and simultaneously to the editors of this book (the CUPS Directors), via electronic mail—see addresses listed below. Please feel free to communicate any suggestions about the programs and text which may lead to improvements in future editions. Since the programs have been provided in source code form, it will be possible for you to make corrections of any errors that you or we find in the future—provided that you send in the registration card at the back of the book, so that you can be notified. The fact that you have the source code will also allow you to make modifications and extensions of the programs. We can assume no responsibility for errors that arise in programs that you have modified. In fact, we strongly urge you to change the program name, and to add a documentary note at the beginning of the code of any modified programs that alerts other potential users of any such changes.

1.6   CUPS Courses and Developers

- **CUPS Directors**
  Maria Dworzecka, George Mason University (cups@gmuvax.gmu.edu)
  Robert Ehrlich, George Mason University (cups@gmuvax.gmu.edu)
  William MacDonald, University of Maryland (w_macdonald@umail.umd.edu)

- **Astrophysics**
  J. M. Anthony Danby, North Carolina State University (n38hs901@ncuvm.ncsu.edu)
  Richard Kouzes, Battelle Pacific Northwest Laboratory (rt_kouzes@pnl.gov)
  Charles Whitney, Harvard University (whitney@cfa.harvard.edu)

- **Classical Mechanics**
  Bruce Hawkins, Smith College (bhawkins@smith.bitnet)
  Randall Jones, Loyola College (rsj@loyvax.bitnet)

- **Electricity and Magnetism**
  Robert Ehrlich, George Mason University (rehrlich@gmuvax.gmu.edu)
  Lyle Roelofs, Haverford College (lroelofs@haverford.edu)
  Ronald Stoner, Bowling Green University (stoner@andy.bgsu.edu)
  Jaroslaw Tuszynski, George Mason University (cups@gmuvax.gmu.edu)

- **Modern Physics**
  Douglas Brandt, Eastern Illinois University (cfdeb@ux1.cts.eiu.edu)
  John Hiller, University of Minnesota, Duluth (jhiller@d.umn.edu)
  Michael Moloney, Rose Hulman Institute (moloney@nextwork.rose-hulman.edu)

- **Nuclear and Particle Physics**
  Roberta Bigelow, Willamette University (rbigelow@willamette.edu)
  John Philpott, Florida State University (philpott@fsunuc.physics.fsu.edu)
  Joseph Rothberg, University of Washington (rothberg@phast.phys.washington.edu)

- **Quantum Mechanics**
  John Hiller, University of Minnesota, Duluth (jhiller@d.umn.edu)
  Ian Johnston, University of Sydney (idj@suphys.physics.su.oz.au)
  Daniel Styer, Oberlin College (dstyer@physics.oberlin.edu)

- **Solid State Physics**
  Graham Keeler, University of Salford (g.j.keeler@sysb.salford.ac.uk)
  Roger Rollins, Ohio University (rollins@chaos.phy.ohiou.edu)
  Steven Spicklemire, University of Indianapolis (steves@truevision.com)

- **Thermal and Statistical Physics**
  Harvey Gould, Clark University (hgould@vax.clarku.edu)
  Lynna Spornick, Johns Hopkins University
  Jan Tobochnik, Kalamazoo College (jant@kzoo.edu)

- **Waves and Optics**
  G. Andrew Antonelli, Wolfgang Christian, and Susan Fischer, Davidson College (wc@phyhost.davidson.edu)
  Robin Giles, Brandon University (giles@brandonu.ca)
  Brian James, Salford University (b.w.james@sysb.salford.ac.uk)

1.7   Descriptions of all CUPS Programs

Each of the computer simulations in this book (as well as those in the eight other books comprised by the CUPS Project) are described below. The individual headings under which programs appear correspond to the nine CUPS courses. In several cases, programs are listed under more than one course. The number of programs listed under the Astrophysics, Modern Physics, and Thermal Physics courses is appreciably greater than the others, because several authors have opted to subdivide their programs into many smaller programs. Detailed inquiries regarding CUPS programs should be sent to the program authors.

ASTROPHYSICS PROGRAMS

STELLAR (Stellar Models), written by Richard Kouzes, is a simulation of the structure of a static star in hydrodynamic equilibrium. This provides a model of a zero age main sequence star, and helps the user understand the physical processes that exist in stars, including how density, temperature, and luminosity depend on mass. Stars are self-gravitating masses of hot gas supported by thermodynamic processes fueled by nuclear fusion at their core. The model integrates the four differential equations governing the physics of the star to reach an equilibrium condition which depends only on the star's mass and composition.

EVOLVE (Stellar Evolution), written by Richard Kouzes, builds on the physics of a static star, and considers (1) how a gas cloud collapses to become a main sequence star, and (2) how a star evolves from the main sequence to its final demise. The model is based on the same physics as the STELLAR program. Starting from a diffuse cloud of gas, a protostar forms as the cloud collapses and reaches a sufficient density for fusion to begin. Once a star reaches equilibrium, it remains for most of its life on the main sequence, evolving off after it has consumed its fuel. The final stages of the star's life are marked by rapid and dramatic evolution.

BINARIES is the driver program for all Binaries programs (**VISUAL1, VISUAL2, ECLIPSE, SPECTRO, TIDAL, ROCHE, and ACCRDISK**).

VISUAL1 (Visual Binaries—Proper Motion), written by Anthony Danby, enables you to visualize the proper motion in the sky of the members of a visual binary system. You can enter the elements of the system and the mass ratio, as well as the speed at which the center of mass moves across the screen. The program also includes an animated three-dimensional demonstration of the elements.

VISUAL2 (Visual Binaries—True Orbit), written by Anthony Danby, enables you to select an apparent orbit for the secondary star with arbitrary eccentricity, with the primary at any interior point. The elements of the orbit are displayed. You can see the orbit animated in three dimensions, or can make up a set of "observations" based on the apparent orbit.

ECLIPSE (Eclipsing Binaries), written by Anthony Danby, shows simultaneously either the light curve and the orbital motion or the light curve and an animation of the eclipses. You can select the elements of the orbit and radii and magnitudes of the stars. A form of limb-darkening is also included as an option.

SPECTRO (Spectroscopic Binaries), written by Anthony Danby, allows you to select the orbital elements of a spectroscopic binary, and then shows simultaneously the velocity curve, the orbital motion, and a moving spectral line.

TIDAL (Tidal Distortion of a Binary), written by Anthony Danby, models the motion of a spherical secondary star around a primary that is tidally distorted by the secondary. You can select orbital elements, masses of the stars, a parameter describing the tidal lag, and the initial rate of rotation of the primary. The equations are integrated over a time interval that you specify. Then you can see the changes of the orbital elements, and the rotation of the primary, with time. You can follow the motion in detail around each revolution, or in a form where the equations have been averaged around each revolution.

ROCHE (The Photo-Gravitational Restricted Problem of Three Bodies), written by Anthony Danby, follows the two-dimensional motion of a particle that is subject to the gravitational attraction of two bodies in mutual circular orbits, and also, optionally, radiation pressure from these bodies. It is intended, in part, as background for the interpretation of the formation of accretion disks. Curves of zero velocity (that limit regions of possible motion) can be seen. The orbits can also be followed using Poincaré maps.

ACCRDISK (Formation of an Accretion Disk), written by Anthony Danby, follows some of the dynamical steps in this process. The dynamics is valid up to the initial formation of a hot spot, and qualititative afterward.

NBMENU is the driver program for all programs on the motion of N interacting bodies: **TWO-GALAX, ASTROIDS, N-BODIES, PLANETS, PLAYBACK, and ELEMENTS**.

TWOGALAX (The Model of Wright and Toomres), written by Anthony Danby, is concerned with the interaction of two galaxies. Each consists of a central gravitationally attracting point, surrounded by rings of stars (which are attracted, but do not attract). Elements of the orbits of one galaxy relative to the other are selected, as is the initial distribution and population of the rings. The motion can be viewed as projected into the plane of the orbit of the galaxies, or simultaneously in that plane and perpendicular to it. The positions can be stored in a file for later viewing.

ASTROIDS (N-Body Application to the Asteroids), written by Anthony Danby, uses the same basic model, but a planet and a star take the place of the galaxies and the asteroids replace the stars. Emphasis is on asteroids all having the same period, with interest on periods having commensurability with the period of the planet. The orbital motion of the system can be followed. The positions can be stored in a file for later viewing. An asteroid can be selected, and the variation of its orbital elements can then be followed.

NBODIES (The Motion of N Attracting Bodies), written by Anthony Danby, allows you to choose the number of bodies (up to 20) and the total energy of the system. Initial conditions are chosen at random, consistent with this energy, and the resulting motion can be observed. During the motion various quantities, such as the kinetic energy, are displayed. The positions can be stored in a file for later viewing.

PLANETS (Make Your Own Solar System), written by Anthony Danby, is similar to the preceding program, but with the bodies interpreted as a star with planets. Initial conditions are specified through the choice of the initial elements of the planets. The positions can be stored in a file for later viewing.

PLAYBACK, written by Anthony Danby, enables a file stored by one of the preceding programs to be viewed.

ELEMENTS (Orbital Elements of a Planet), written by Anthony Danby, shows a three-dimensional animation that can be viewed from any angle.

GALAXIES is the driver program for Galactic Kinematics programs: **ROTATION, OORTCONS, and ARMS21CM**.

ROTATION (The Rotation Curve of a Galaxy), written by Anthony Danby, first prompts you to "design" a galaxy, consisting of a central mass and up to five spheroids (that can be visible or invisible). It then displays the galaxy and can show the animated rotation or the rotation curve.

OORTCONS (Galactic Kinematics and Oort's Constants), written by Anthony Danby, allows you to design your galaxy, choose the location of the "sun" and a local region around it, and the to observe the kinematics in this region. It also shows graphs of radial velocity and proper motion in comparison with the linear approximation, and computes the Oort constants.

ARMS21CM (The Spiral Structure of a Galaxy), written by Anthony Danby, allows you to design your galaxy, construct a set of spiral arms, and select the position of the "sun." Then, for different galactic longitudes, you can see observed profiles of 21 cm lines.

ATMOS (Stellar Atmospheres), written by Charles Whitney, permits the user to select a constellation, see it mapped on the computer screen, point to a star, and see it plotted on a brightness-color diagram. The user's task is to build a model atmosphere that imitates the photometric properties of observed stars. This is done by specifying numerical values for three basic stellar parameters: radius, mass, and luminosity. The program then builds the model and displays it on the brightness-color diagram, and it also plots the spectrum and the detailed thermodynamic structure of the atmosphere. With this program the user may investigate the relation between stellar parameters and the thermal properties of the gas in the atmosphere. Two atmospheres may be superposed on the graphs, for easier comparison.

PULSE (Stellar Pulsations), written by Charles Whitney, illustrates stellar pulsation by simulating the thermo-mechanical behavior of a "star" modeled by a self-gravitating gas divided by spherical elastic shells. The elastic shells resemble a set of coupled oscillators. The program solves for the modes of small-amplitude motion, and it uses Fourier synthesis to construct motions for arbitrary starting conditions. The screen displays the thermodynamic structure and surface properties, such as temperature, pressure, and velocity. Animation displays the nature of the pulsation. By showing the motions, temperatures, and energy flux, the program demonstrates the heat engine acting inside the pulsating star. The motions of the shells and the spatial Fourier decomposition into eigenmodes are displayed simultaneously, and this will help you visualize the meaning of the Fourier components.

CLASSICAL MECHANICS PROGRAMS

GENMOT (The Motion Generator), written by Randall Jones, allows you to solve numerically any differential equation of motion for a system with up to three degrees of freedom and display the time evolution of the system in a wide variety of formats. Any of the dynamical variables or any function of those variables may be displayed graphically and/or numerically and a wide range of animations may be constructed. Since the Motion Generator can be used to solve any second-order differential equation, it can also be used to study systems analyzed by Lagrangian methods. Real world coordinates may be constructed as functions of generalized coordinates so that simulations of the actual system can be constructed.

ROTATE (Rotation of 3-D Objects), written by Randall Jones, is designed to aid in the visualization of the dynamical variables of rotational motion. It will allow you to observe the 3-D motion of rotating objects in a controlled fashion, running the simulation faster, slower, or in reverse while displaying the corresponding evolution of the angular velocity, the angular momentum and the torque. It will display the motion from the fixed frame and from the body frame to help in understanding the translation between these two descriptions of the motion. By using the stereographic feature of the program you can create a genuine 3-D representation of the motion of the quantities.

COUPOSC (Coupled Oscillators), written by Randall Jones, is designed to investigate a wide range of harmonic systems. Given a set of objects and springs connected in one or two dimensions, the simulation can solve the problem by generating the normal mode frequencies and their corresponding motions. It can take any set of initial conditions and resolve them into their component normal mode motions or take any set of initial mode occupations and display the corresponding motions of the objects. It can also determine the motion of the system when it is acted on by external forces. In this case the total forces are no longer harmonic, so the solution is generated numerically. The harmonic analysis, however, still provides an important tool for investigating and understanding the subsequent motion.

ANHARM (Anharmonic Oscillators), written by Bruce Hawkins, simulates oscillations of various types: pendulum, simple harmonic oscillator, asymmetric, cubic, Vanderpol, and a mass in the center of a spring with fixed ends. Nonlinear behavior is emphasized. The user may choose to view one to four graphs of the motion simultaneously, along with the potential diagram and a picture of the moving object. Graphs that may be viewed are x vs. t, v vs. t, v vs. x, the Poincaré diagram, and the return map. Tools are provided to explore parameter space for regions of interest. Fourier analysis is available, resonance diagrams can be plotted, and the period can be plotted as a function of energy. Includes a tutorial demonstrating the usefulness of phase plots and Poincaré plots.

ORBITER (Gravitational Orbits), written by Bruce Hawkins, simulates the motion of up to five objects with mutually gravitational attraction, and any reasonable number of additional objects moving in the gravitation field of the first five. The motion may be viewed in up to six windows simultaneously: windows centered on a particular body, on the center of mass, stationary in the universe frame, or rotating with the line joining the two most massive bodies. A menu of available systems includes the solar system, the sun/earth/moon system; the sun, Jupiter, and its moons; the sun, earth, and Saturn, demonstrating retrograde motion; the sun, Jupiter, and a comet; and a pair of binary stars with a comet. Bodies may be added to any system, or a new system created using either numerical coordinates or the mouse. Bodies may be replicated to demonstrate the sensitivity of orbits to initial conditions.

COLISION (Collisions), written by Bruce Hawkins, simulates two-body collisions under any of a number of force laws: Coulomb with and without shielding and truncation, hard sphere, soft sphere (harmonic), Yukawa, and Woods-Saxon. Collision may be viewed in the laboratory and center of mass systems simultaneously, with or without momentum diagrams. Includes a tutorial on the usefulness of the center of mass system, one on the kinematics of relativistic collisions, and one on cross section. Plots cross section against scattering parameter, and compares collisions at different parameters.

ELECTRICITY AND MAGNETISM PROGRAMS

FIELDS (Analysis of Vector and Scalar Fields), written by Jarek Tuszynski, displays scalar and vector fields for any algebraic or trigonometric expression entered by the user. It also computes numerically the divergence, curl, and Laplacian for the vector fields, and the gradient and Laplacian for the scalar fields. Simultaneous displays of selected quantities are provided in user-selected planes, using vector, contour, or 3-D plots. The program also allows the user to define paths along which line integrals are computed.

GAUSS (Gauss' Law), written by Jarek Tuszynski, treats continuous charge distributions having spherical or cylindrical symmetry, and those that vary as a function of the x-coordinate only. The program allows the user to enter an arbitrary function to define either the electric field magnitude, the potential, or the charge density. It then computes the other two functions by numerical differentiation or integration, and displays all three functions. Finally, the program allows the user to enter a "comparison function," which is plotted on the same graph, so as to check whether his analytic solutions are correct.

POISSON (Poisson's Equation Solved on a Grid), written by Jarek Tuszynski, solves Poisson's equation iteratively on a 2-D grid using the method of simultaneous over-relaxation. The user can draw arbitrary systems consisting of line charges, and charged conducting cylinders, plates, and wires, all infinite in extent perpendicular to the grid. After iteratively solving Poisson's equation, the program displays the results for the potential, electric field, or the charge density (found from the Laplacian of the potential), in the form of contour, vector, or 3-D plots. In addition, many other program features are available, including the ability to specify surfaces, along which the potential varies according to some algebraic function specified by the user.

IMAG&MUL (Image Charges and Multipole Expansion), written by Lyle Roelofs and Nathaniel Johnson, allows users to explore two approaches to the solution of Laplace's equation—the image charge method and expansion in multipole moments. In the image charge mode (IC) the user is presented with a variety of configurations involving conducting planes and point charges and is asked to "solve" each by placing image charges in the appropriate locations. The program displays the electric field due to all point charges, real and image, and a solution can be regarded as successful with the field due to all charges is everywhere orthogonal to all conducting surfaces. Solutions can then be examined with a variety of included software "tools." The multipole expansion (ME) mode of the program also permits a "hands-on" exploration of standard electrostatic problems, in this case the "exterior" problem, i.e., the determination of the field outside a specified equipotential surface. The program presents the user with a variety of azimuthally symmetric equipotential surfaces. The user "solves" for the full potential by adding chosen amounts of the (first six) multipole moments. The screen shows the contours of the summed potential and the problem is "solved" when the innermost contour matches the given equipotential surface as closely as possible.

ATOMPOL (Atomic Polarization), written by Lyle Roelofs and Nathaniel Johnson, is an exploration of the phenomenon of atomic polarization. Up to 36 atoms of controllable polarizability are immersed in an external electric field. The program solves for and displays the field throughout the region in which the atoms are located. A closeup window shows the polarization of selected atoms and software "tools" allow for further analysis of the resulting electric fields. Use of this program improves the student's understanding of polarization, the interaction of polarized entities, and the atomic origin of macroscopic polarization, the latter via study of closely spaced clusters of polarizable atoms.

DIELECT (Dielectric Materials), written by Lyle Roelofs and Nathaniel Johnson, is a simulation of the behavior of linear dielectric materials using a cell-based approach. The user controls either the polarization or the susceptibility of each cell in a (25×25) grid (with assumed uniformity in the third direction). Full self-consistent solutions are obtained via an iterative relaxation method and the fields P, E, or D are displayed. The student can investigate the self-interactions of polarized materials and many geometrical effects. Use of this program aids the student in developing understanding of the subtle relations among and meaning of P, E, and D.

ACCELQ (Fields From an Accelerated Charge), written by Ronald Stoner, simulates the electromagnetic fields in the plane of motion generated by a point charge that is moving and accelerating in two dimensions. The user chooses from among seven predefined trajectories, and sets the values of maximum speed and viewing time. The electric field pattern is recomputed after each change of trajectory or parameter; thereafter, the user can investigate the electric field, magnetic field, retarded potentials, and Poynting-vector field by using the mouse as a field probe, by using gridded overlays, or by generating plots of the various fields along cuts through the viewing plane.

QANIMATE (Fields From an Accelerated Charge—Animated Version), written by Ronald Stoner, is an interactive animation of the changing electric field pattern generated by a point electric charge moving in two dimensions. Charge motion can be manipulated by the user from the keyboard. The display can include electric field lines, radiation wave fronts, and their points of intersection. The motion of the charge is controlled by the using **arrow** keys to accelerate and steer much like the accelerator and steering wheel of a car, except that acceleration must be changed in increments, and the **Space** bar can used to engage or disengage the steering. With steering engaged, the charge will move in a circle. Unless the acceleration is made zero, the speed will increase (or decrease) to the maximum (minimum) possible value. At constant speed and turning rate, the charge can be controlled by the **Space** bar alone.

EMWAVE (Electromagnetic Waves), written by Ronald Stoner, uses animation to illustrate the behavior of electric and magnetic fields in a polarized plane electromagnetic wave. The user can choose to observe the wave in free space, or to see the effect on the wave of incidence on a material interface, or to see the effects of optical elements that change its polarization. The user can change the polarization state of the incident wave by specifying its Stokes parameters. Standing electromagnetic waves can be simulated by combining the incident traveling wave with a reflected wave of the same amplitude. The user can do that by choosing appropriate values of the physical properties of the medium on which the incident wave impinges in one of the animations.

MAGSTAT (Magnetostatics), written by Ronald Stoner, computes and displays magnetic fields in and near magnetized materials. The materials are uniform and have 3-D shapes that are solids of revolution about a vertical axis. The shape of the material can be modified or chosen from a data input screen. The user has the option of generating the fields produced by a permanently and uniformly magnetized object, or of generating the fields of a magnetizable object placed in an otherwise uniform external field. Besides choosing the shape and aspect ratio of the object, the user can vary the magnetic permeability of the magnetizable material, and choose among three fields to display: magnetic induction (B), magnetic field strength (H), and magnetization (M). Each of these fields can be displayed or explored in several different ways. The algorithm for computing the fields uses a superposition of Chebyschev polynomial approximants to the H field due to "rings" of "magnetic charge."

MODERN PHYSICS PROGRAMS

NUCLEAR (Nuclear Energetics and Nuclear Counting), written by Michael Moloney, deals with basic nuclear properties related to mass, charge, and energy, for approximately 1900 nuclides. Graphs are available involving binding energy, mass, and Q values of a variety of nuclear reactions, including alpha and beta decays. Part 2 deals with simulating the statistics of counting with a Geiger-Muller tube. This part also simulates neutron activation, and the counting behavior as neutron flux is turned on and off. Finally, a decay chain from A to B to C is simulated, where half-lives may be changed, and populations are graphed as a function of time.

GERMER (Davisson-Germer and G. P. Thomson Experiments), written by Michael Moloney, simulates both the Davisson-Germer and G. P. Thomson experiments with electrons scattering from crystalline materials. Stress is laid on the behavior of electrons as waves; similarities are noted with scattering of x-rays. The exercises encourage students to understand why peaks and valleys in scattered electrons occur where they do.

QUANTUM (one-dimensional Quantum Mechanics), written by Douglas Brandt, is a program that has four sections. The first section allows users to investigate the uncertainty principle for specified wavefunctions in position or momentum space. The second section allows users to investigate the time evolution of wavepackets under various dispersion relations. The third section allows users to investigate solutions to Schrödinger's equation for asymptotically free solutions. The user can input a barrier and the program calculates reflection and transmission coefficients for a range of energies and show wavepacket time evolution for the barrier potential. The fourth section is similar to the third, except that it allows the user to investigate bound solutions to Schrödinger's equation. The program calculates the bound state Hamiltonian eigenvalues and spatial eigenfunctions.

RUTHERFD (Rutherford Scattering), written by Douglas Brandt, is a program for investigating classical scattering of particles. A scattering potential can be chosen from a list of predefined potentials or an arbitrary potential can be input by the user. The computer generates scattering events by randomly picking impact parameters from a distribution defined by beam parameters specified by the user. It displays the results of the scattering on a polar histogram and on a detailed histogram to help users gain insight into differential scattering cross section. A scintillation mode can be chosen for users that want more appreciation of the actual experiments of Geiger and Marsden. A "guess the scatterer" mode is available for trying to gain appreciation of how scattering experiments are used to infer properties of the scatterers.

SPECREL (Special Relativity), written by Douglas Brandt, is a program to investigate special relativity. The first section is to investigate change of coordinate systems through Minkowski diagrams. The user can define coordinates of objects in one reference frame and the computer calculates the coordinates in a user-selectable coordinate system and displays the objects in both reference frames. The second section allows users to view clocks that are in relative motion. A clock can be given an arbitrary trajectory through space-time and the readings of various clocks can be viewed as the clock follows that trajectory. A third section allows users to observe collisions in different reference frames that are related by Lorentz transformations or by Gallilean transformations.

LASER (Lasers), written by Michael Moloney, simulates a three-level laser, with the user in control of energy level parameters, temperature, pump power, and end mirror transmission. Atomic populations may be graphically tracked from thermal equilibrium through the lasing threshold. A mirror cavity simulation is available which uses ray tracing. This permits study of cavity stability as a function of mirror shape and position, as well as beam shape characteristics within the cavity.

HATOM (Hydrogenic Atoms), written by John Hiller, computes eigenfunctions and eigenenergies for hydrogen, hydrogenic atoms, and single-electron diatomic ions. Hydrogenic atoms may be exposed to uniform electric and magnetic fields. Spin interactions are not included. The magnetic interaction used is the quadratic Zeeman term; in the absence of spin-orbit coupling, the linear term adds only a trivial energy shift. The unperturbed hydrogenic eigenfunctions are computed directly from the known solutions. When external fields are included, approximate results are obtained from basis-function expansions or from Lanczos diagonalization. In the diatomic case, an effective nuclear potential is recorded for use in calculation of the nuclear binding energy.

NUCLEAR AND PARTICLE PHYSICS PROGRAMS

NUCLEAR (Nuclear Energetics and Counting), written by Michael Moloney, is included here, but is described under the Modern Physics heading.

SHELLMOD (Nuclear Models), written by Roberta Bigelow, calculates energy levels for spherical and deformed nuclei using the single particle shell model. You can explore how the nuclear potential shape, the spin-orbit interaction, and deformation affect both the order and spacing of nuclear energy levels. In addition, you will learn how to predict spin and parity for single particle states.

NUCRAD (Interaction of Radiation With Matter), written by Roberta Bigelow, is a simulation of alpha particles, muons, electrons, or photons interacting with matter. You will develop an understanding of how ranges, energy losses, and random particle paths depend on materials, radiation, and incident energy. As a specific application, you can explore photon and electron interactions in a sodium iodide crystal which determines the energy response of a radiation detector.

ELSCATT (Electron-Nucleus Scattering), by John Philpott, is an interactive software tool that demonstrates various aspects of electron scattering from nuclei. Specific features include the relativistic kinematics of electron scattering, densities and form factors for elastic and inelastic scattering, and the nuclear Coulomb response. The simulation illustrates how detailed nuclear structure information can be obtained from electron scattering measurements.

TWOBODY (Two-Nucleon Interactions), by John Philpott, is an interactive software tool that illuminates many features of the two-nucleon problem. Bound state wavefunctions and properties can be calculated for a variety of interactions that may include non-central parts. Phase shifts and cross sections for pp, pn, and nn scattering can be calculated and compared with those obtained experimentally. Spin-polarization features of the cross sections can be extensively investigated. The simulation demonstrates the richness of the two-nucleon data and its relation to the underlying nucleon-nucleon interaction.

RELKIN (Relativistic Kinematics), by Joseph Rothberg, is an interactive program to permit you to explore the relativistic kinematics of scattering reactions and two-body particle decays. You may choose from among a large number of initial and final states. The initial momentum of the beam particle and the center of mass angle of a secondary can also be specified. The program displays the final state vector momenta in both the lab system and center of mass system along with numerical values of the most important kinematic quantities. The program may be run in a Monte Carlo mode, displaying a scatter plot and histogram of selected variables. The particle data base may be modified by the user and additional reactions and decay modes may be added.

DETSIM (Particle Detector Simulation), by Joseph Rothberg, is an interactive tool to allow you to explore methods of determining parameters of a decaying particle or scattering reaction. The program simulates the response of high-energy particle detectors to the final-state particles from scattering or decays. The detector size and location may be specified by the user as well as its energy and spatial resolution. If the program is run in a Monte Carlo mode, detector hit information for each event is written to a file. This file can be read by a small reconstruction and plotting program. You can easily modify one of the example reconstruction programs that are provided to determine the mass, momentum, and other properties of the initial particle or state.

QUANTUM MECHANICS PROGRAMS

BOUND1D (Bound States in One Dimension), written by Ian Johnston, is a tool which allows you to explore energy eigenfunctions for an electron in various potential wells, which can be square, parabolic, ramped, asymmetric, double, or Coulombic. The first part of the program deals with finding the eigenvalues and eigenfunctions of different wells. You may find them yourself, using a "hunt and shoot" method, or else the program will compute the eigenvalues automatically, by counting the number of nodes to determine where the eigenvalues occur. The second part of the program looks at properties of eigenfunctions normalization, orthogonality, and the evaluation of many kinds of overlap integrals. The third part examines the time development of general states made up of a superposition of bound state eigenfunctions. Facility is provided for you to incorporate your own procedures to specify different potential wells or different overlap integrals.

SCATTR1D (Scattering in One Dimension), written by John Hiller, solves the time-independent Schrödinger equation for stationary scattering states in one-dimensional potentials. The wavefunction is displayed in a variety of ways, and the transmission and reflection probabilities are computed. The probabilities may be displayed as functions of energy. The computations are done by numerically integrating the Schrödinger equation from the region of the transmitted wave, where the wavefunction is known up to some overall normalization and phase, to the region of the incident wave. There the reflected and incident waves are separated. The potential is assumed to be zero in the incident region and constant in the transmitted region.

QMTIME (Quantum Mechanical Time Development), written by Daniel Styer, simulates quantal time development in one dimension. A variety of initial wave packets (Gaussian, Lorentzian, etc.) can evolve in time under the influence of a variety of potential energy functions (step, ramp, square well, harmonic oscillator, etc.) with or without an external driving force. A novel visualization technique simultaneously displays the magnitude and phase of complex-valued wave functions. Either position-space or momentum-space wave functions, or both, can be shown. The program is particularly effective in demonstrating the classical limit of quantum mechanics.

LATCE1D (Wavefunctions on a one-dimensional Lattice), written by Ian Johnston, is a tool which allows you to explore energy eigenfunctions for an electron in a lattice made up of a number of simple potential wells (up to twelve), which can be square, parabolic, or Coulombic. You may find the eigenvalues yourself, using a "hunt and shoot" method, or allow the program to compute them automatically. You can firstly explore regular lattices, where all wells are the same and spaced at regular intervals. These will demonstrate many of the properties of regular crystals, particularly the existence of energy bands. Secondly you can change the width, depth or spacing of any of the wells, which will mimic the effect of impurities or other irregularities in a crystal. Lastly you can apply an external electric across the lattice. Facility is provided for you to incorporate your own procedures to calculate wells, lattice arrangements or external fields of their own choosing.

BOUND3D (Bound States in Three Dimensions), written by Ian Johnston, is a tool which allows you to explore energy eigenfunctions for a particle in a spherically symmetric potential well, which can be square, parabolic, Coulombic, or several other shapes of importance in molecular or nuclear applications. The first part of the program deals with finding the eigenvalues and eigenfunctions of different wells, assuming that the angular part of the wavefunctions are spherical harmonics. You may find them yourself for a given angular momentum quantum number using a "hunt and shoot" method, or else the program will compute the eigenvalues automatically, by counting the number of nodes to determine where the eigenvalues occur. The second part of the program looks at properties of eigenfunctions normalization, orthogonality, and the evaluation of many kinds of overlap integrals. Facility is provided for you to incorporate your own procedures to specify different potential wells or different overlap integrals.

IDENT (Identical Particles in Quantum Mechanics), written by Daniel Styer, shows the probability density associated with the symmetrized, antisymmetrized, or nonsymmetrized wave functions of two noninteracting particles moving in a one-dimensional infinite square well. It is particularly valuable for demonstrating the effective interaction of noninteracting identical particles due to interchange symmetry requirements.

SCATTR3D (Scattering in Three Dimensions), written by John Hiller, performs a partial-wave analysis of scattering from a spherically symmetric potential. Radial and 3-D wavefunctions are displayed, as are phase shifts, and differential and total cross sections. The analysis employs an expansion in the natural angular momentum basis for the scattering wavefunction. The radial wavefunctions are computed numerically; outside the region where the potential is important they reduce to a linear combination of Bessel functions which asymptotically differs from the free radial wavefunction by only a phase. Knowledge of these phase shifts for the dominant values of angular momentum is used to approximate the cross sections.

CYLSYM (Cylindrically Symmetric Potentials), written by John Hiller, solves the time-independent Schrödinger equation Hu=Eu in the case of a cylindrically symmetric potential for the lowest state of a chosen parity and magnetic quantum number. The method of solution is based on evolution in imaginary time, which converges to the state of the lowest energy that has the symmetry of the initial guess. The Alternating Direction Implicit method is used to solve a diffusion equation given by $HU = -\hbar\partial U/\partial t$, where H is the Hamiltonian that appears in the Schrödinger equation. At large times, U is nearly proportional to the lowest eigenfunction of H, and the expectation value $\langle H \rangle = \langle U|H|U \rangle / \langle U|U \rangle$ is an estimate for the associated eigenenergy.

SOLID STATE PHYSICS

LATCE1D (Wavefunctions for a one-dimensional Lattice), written by Ian Johnston, and included here, is described under the Quantum Mechanics heading.

SOLIDLAB (Build Your Own Solid State Devices), written by Steven Spicklemire, is a simulation of a semiconductor device. The device can be "drawn" by the user, and the characteristics of the device adjusted by the user during the simulation. The user can see how charge density, current density, and electric potential vary throughout the device during its operation.

LCAOWORK (Wavefunctions in the LCAO Approximation), written by Steven Spicklemire, is a simulation of the interaction of 2-D atoms within small atomic clusters. The atoms can be adjusted and moved around while their quantum mechanical wavefunctions are calculated in real time. The student can investigate the dependence of various properties of these atomic clusters on the properties of individual atoms, and the geometric arrangement of the atoms within the cluster.

PHONON (Phonons and Density of States), written by Graham Keeler, calculates and displays phonon dispersion curves and the density of states for a number of different 3-D cubic crystal structures. The displays of the dispersion curves show realistic curves and allow the user to study the effect of changing the interatomic forces between nearest and further neighbor atoms and, for diatomic crystal structures, changing the ratio of the atomic masses. The density of states calculation shows how the complex shapes of real densities of states are built up from simpler distributions for each mode of polarization, and enables the user to match the features of the distribution to corresponding features on the dispersion curves. In order to help with visualization of the crystal lattices involved, the program also shows 3-D projections of the different crystal structures.

SPHEAT (Calculation of Specific Heat), written by Graham Keeler, calculates and displays the temperature variation of the lattice specific heat for a number of different theoretical models, including the Einstein model and the Debye model. It also makes the calculation for a computer simulation of a realistic density of states, in which the user can vary the important parameters of the crystal, including those affecting the density of states. The program can display the results for a small region near the origin, and as a T-cubed plot to enable the user to investigate the low temperature limit of the specific heat, or in the form of the equivalent Debye temperature to enhance a study of the deviations from the Debye model. The Schottky specific heat anomaly can also be investigated.

BANDS (Energy Bands), written by Roger Rollins, calculates and displays, for easy comparison, the energy dispersion curves and corresponding wavefunctions for an electron in a 1-D symmetric $V(x) = V(-x)$ periodic potential of arbitrary shape and of strength V_0. The method used is based on an exact, non-perturbative approach so that the energy dispersion curves and band gaps can be obtained for large V_0. Wavefunctions can be displayed, and compared with one another, by clicking the mouse on the desired states on the energy dispersion curve. Changes in band structure can be followed as changes are made in the shape of the potential. The variation of the band gaps with V_0 is calculated and compared with the two opposite limits of very weak V_0 (perturbation method) and very strong V_0 (isolated atom). Even the experienced condensed matter researcher may be surprised by some of the results! Open-ended class discussions can result from the interesting physics found in these conceptually simple model calculations.

PACKET (Electron Wavepacket in a 1-D Lattice), written by Roger Rollins, shows a live animation, calculated in real time, demonstrating how an electron wavepacket in a metal or semiconducting crystal moves under the influence of external forces. The time-dependent Schrödinger equation is solved in a tight binding approximation, including the external force terms, and the motion of the wavepacket is obtained directly. The main objective of the simulation is to show that an electron wavepacket formed from states with energies near the top of an energy band is accelerated in a direction *opposite* to the direction of the external force; it has a *negative* effective mass! The simulation deals with motion in a 1-D lattice but the concepts are applicable to the full 3-D motion of an electron in a real crystal. Numerical experiments on the motion of the packet explore interesting physics questions such as: how does constant applied force affect the periodic motion of a packet? when does the usual semiclassical model fail? what happens to the dynamics of the packet when placed in a superlattice with lattice constant twice that of the original lattice?

THERMAL AND STATISTICAL PHYSICS PROGRAMS

ENGDRV, written by Lynna Spornick, is a driver program for **ENGINE, DIESEL, OTTO, and WANKEL**. These programs provide an introduction to the thermodynamics of engines.

ENGINE (Design Your Own Engine), written by Lynna Spornick, lets the user design an engine by specifying the processes (adiabatic, isobaric, isochoric [constant volume], and isothermic) in the engine's cycle, the engine type (reversible or irreversible), and the gas type (helium, argon, nitrogen, or steam). The thermodynamic properties (heat exchanged, work done, and change in internal energy) for each process and the engine's efficiency are computed.

DIESEL, OTTO, and WANKEL, written by Lynna Spornick, provide animations of each of these types of engine. Plots of the temperature versus entropy and the pressure versus volume for the cycles are shown with the engine's current thermodynamic conditions indicated.

PROBDRV, written by Lynna Spornick, is a driver program for **GALTON, POISEXP, TWOD, KAC, and STADIUM**. Subprograms GALTON, POISEXP, and TWOD provide an introduction to probability and subprograms KAC and STADIUM provide an introduction to statistics.

GALTON (A Galton Board), written by Lynna Spornick, models either a traditional Galton Board or a customized Galton Board with traps, reflecting, and/or absorbing walls. GALTON demonstrates the binominal and normal distributions, the laws of probability, and the central limit theorem.

POISEXP (Poisson Probability Distribution in Nuclear Decay), written by Lynna Spornick, uses the decay of radioactive atoms to describe the Poisson and the exponential distributions.

TWOD (2-D Random Walk), written by Lynna Spornick, models a random walk in two dimensions. A "drunk," taking equal-length steps, is required to walk either on a grid or on a plane. TWOD demonstrates the joint probability of two independent processes, the binominal distribution, and the Rayleigh distribution.

KAC (A Kac Ring), written by Lynna Spornick, uses a Kac ring to demonstrate that large mechanical systems, whose equations of motion are solvable and which obey time reversal and have a Poincaré cycle, can also be described by statistical models.

STADIUM (The Stadium Model), written by Lynna Spornick, uses a stadium model to demonstrate that there exist mechanical systems whose equations of motion are solvable but whose motion is not predictable because of the system's chaotic nature.

ISING (Ising Model in One and Two Dimensions), written by Harvey Gould, allows the user to explore the static and dynamic properties of the 1- and 2-D Ising model using four different Monte Carlo algorithms and three different ensembles. The choice of the Metropolis algorithm allows the user to study the Ising model at constant temperature and external magnetic field. The orientation of the spins is shown on the screen as well as the evolution of the total energy or magnetization. The mean energy, magnetization, heat capacity, and susceptibility are monitored as a function of the number of configurations that are sampled. Other computed quantities include the equilibrium-averaged energy and magnetization autocorrelation functions and the energy histogram. Important physical concepts that can be studied with the aid of the program include the Boltzmann probability, the qualitative behavior of systems near critical points, critical exponents, the renormalization group, and critical slowing down. Other algorithms that can be chosen by the user correspond to spin exchange dynamics (constant magnetization), constant energy (the demon algorithm), and single cluster Wolff dynamics. The latter is particularly useful for generating equilibrium configurations at the critical point.

MANYPART (Many Particle Molecular Dynamics), written by Harvey Gould, allows the user to simulate a dense gas, liquid, or solid in two dimensions using either molecular dynamics (constant energy, constant volume) or Monte Carlo (constant temperature, constant volume) methods. Both hard disks and the Lennard-Jones interaction can be chosen. The trajectories of the particles are shown as the system evolves. Physical quantities of interest that are monitored include the pressure, temperature, heat capacity, mean square displacement, distribution of the speeds and velocities, and the pair correlation function. Important physical concepts that can be studied with the aid of the program include the Maxwell-Boltzmann probability distribution, fluctuations, equation of state, correlations, and the importance of chaotic mixing.

FLUIDS (Thermodynamics of Fluids), written by Jan Tobochnik, allows the user to explore the fluid (gas and liquid) phase diagrams for the van der Waals model and water. The user chooses four phase diagrams from among the following choices: *PT, Pv, vT, uT, sT, uv,* and *sv,* where *P* is the pressure, *T* is the temperature, *v* is the specific volume, *S* is the specific entropy, and *u* is the specific internal energy. The program reads in the coexistence table for the van der Waals model and water, and uses it along with an empirical formula for the water free energy and the free energy derived from the van der Waals model. Given v and u, any thermodynamic quantity can be calculated. For the van der waals model thermodynamic quantities also can be calculated from the other thermodynamic state variables. The user can draw a straight line path in one phase diagram and see how this path looks in the other phase diagrams. The user also can extract all important thermodynamic data at any point in a phase diagram.

QMGAS1 (Quantum Mechanical Gas—Part 1), written by Jan Tobochnik, does the numerical calculations necessary to solve for the thermodynamic properties of quantum ideal gases, including photons in blackbody radiation, ideal bosons, phonons in the Debye theory, non-interacting fermions, and the classical limits of these systems. The user chooses the type of statistics (Bose-Einstein, Fermi-Dirac, or Maxwell-Boltzmann), the dimension of space, the form of the dispersion relation (restricted to simple powers), whether or not the particles have a non-zero chemical potential, and whether or not there is a Debye cutoff. The program then allows the user to build up a table of thermodynamic data, including the energy, specific heat, and chemical potential as a function of temperature. This data and various distribution functions and the density of states can be plotted.

QMGAS2 (Quantum Mechanical Gas—Part2), written by Jan Tobochnik, implements a Monte Carlo simulation of a finite number of quantum particles fluctuating between various states in a finite k-space (k is the wavevector). The program orders the possible energy states into an energy level diagram and then allows particles to move from one state to another according to the usual Boltzman probability distribution. Bosons are restricted so that they may not pass through each other on the energy level diagram; fermions are further restricted so that no two fermions may be in the same state; classical particles have no restrictions. In this way indistinguishability is correctly implemented for bosons and fermions. The user chooses the type of particle, the number of particles, the size and dimension of k-space, and the temperature. During the simulation the user sees a representation of the state occupancy and plots of the average energy, the instantaneous energy, and the distribution of energy among the states, also shown are results for the average energy, specific heat, and the occupancy of the ground state.

WAVES AND OPTICS PROGRAMS

DIFFRACT (Interference and Diffraction), by Robin Giles, simulates some of the fundamental wave behaviors in Fresnel and Fraunhofer Diffraction, and other Interference and Coherence effects. In particular you will be able to study diffraction phenomena associated with a point or a set of points and a slit or set of slits using the Huyghens construction. You can also use a method developed by Cornu—the Cornu Spiral—to examine diffraction from one or two slits or one or two obstacles. You can study Fresnel and Fraunhofer diffraction with a single slit or set of slits, a rectangular aperture and a circular aperture. Finally you can study Partial Coherence and fringe visibility in interference and diffraction observations. In the latter example you will be able to study the Michelson Stellar Interferometer and measure the separation distance in a double star and measure the diameter of single stars.

SPECTRUM (Applications of Interfence and Diffraction), by Robin Giles, simulates the uses and modes of operation of four important optical instruments—the Diffraction Grating, the Prism Spectrometer, the Michelson Interferometer and the Fabry-Perot Interferometer. You will look at the nature of the spectra, simulated interference patterns, and the question of resolving power.

WAVE (One-Dimensional Waves), by Wolfgang Christian, Andrew Antonelli, and Susan Fischer, uses finite difference methods to study the time evolution of the following partial differential equations: classical wave, Schrödinger, diffusion, Klein-Gordon, sine-Gordon, phi four, and double sine-Gordon. The user may vary the initial function and boundary conditions. Unique features of the program include mouse-driven drawing tools that enable the user to create sources, segments, and detectors anywhere inside the medium. Double-clicking on a segment allows the user to edit properties such as index of refraction or potential in order to model barrier problems such as thin film interference filters or the Ramsauer-Townsend effect in optics and quantum mechanics, respectively. Various types of analysis can be performed, including detector value, space-time, Fourier analysis and energy density.

CHAIN (One-Dimensional Lattice of Coupled Oscillators), by Wolfgang Christian, Andrew Antonelli, and Susan Fischer, allows the user to examine the time evolution of a 1-D lattice of coupled oscillators. These oscillators are represented on screen as a chain of masses undergoing vertical displacement. The program allows the user to examine how the application of Newtonian mechanics to these masses leads to traveling and standing waves. The relationship between the lattice spacing and other properties such as dispersion, band gaps, and cut-off frequency can be examined. Each mass can be assigned linear, quadratic, and cubic nearest neighbor interactions as well as a time-dependent external force function. Global properties such as the total energy in the lattice or the Fourier transform of the lattice can be displayed as well as the time evolution of a single mass's dynamical variables.

FOURIER (Fourier Analysis and Synthesis), written by Brian James, allows investigation of Fourier analysis and 1-D and 2-D Fourier transforms. In Fourier analysis users can choose from several predefined functions or enter their own functions either algebraically, numerically, or graphically. The build-up of a periodic function is illustrated as successive terms of the Fourier series are added in, and the effects of dispersion and attenuation on the evolution of the synthesized waveform can then be investigated. One- and two-dimensional discrete Fourier transforms can be produced for a range of standard and user-entered functions. The effects of filters on the inverse transforms are illustrated. The 2-D transforms are shown as surface and contour plots. Image processing can be illustrated by filtering the transforms of gray level images so that when the inverse transforms are displayed it can be seen that the images have been modified.

RAYTRACE (Ray Tracing and Lenses), by Brian James, lets the user explore the applications of ray tracing in geometrical optics. The fundamental principle of Fermat can be illustrated by plotting the path of a ray through two different materials between fixed points. The variation of the path of a ray through a region of changing refractive index can be used to investigate the formation of mirages. The variation of pulse delay in a fiber can be calculated as a function of its parameters and the characteristics of optical communication fibers are considered. The formation of primary and secondary rainbows due to dispersion of refractive index can be displayed. The matrix method of tracing rays through lenses can be used to investigate the images formed and show how aberrations in images arise and may be reduced.

QUICKRAY (Quick Ray Tracing), by John Philpott, can be used to demonstrate ray diagrams for a single thin lens or spherical mirror. The object and image are shown, along with the three principal rays that proceed from the object towards the observer. You can use the mouse to move the object, the position of the lens or mirror or to change the focal length of the lens or mirror. The principal rays are continuously redrawn while any of these adjustments are made. The simulation handles converging and diverging lenses and concave and convex mirrors. Thus students can quickly get an intuitive feel for real and virtual image formation under a variety of circumstances.

Acknowledgments

The CUPS Project was funded by the National Science Foundation (under grant number PHY-9014548), and it has received support from the IBM Corporation, the Apple Corporation, and George Mason University.

2

Binary Stars

J.M. Anthony Danby

2.1 Introduction

Observation suggests that in the solar neighborhood most stars are members of gravitationally bound multiple systems. In regions of high stellar density, such as a globular cluster or the center of a galaxy, multiple stellar systems are probably disrupted due to encounters with other stars; but, in general, multiplicity seems to be the rule. Here we are concerned strictly with gravitationally bound binary systems, consisting of just two stars. This is not unduly restrictive; in a triple system, the distance of one star to the center of mass of the others will be at least an order of magnitude greater than the latters' separation. To investigate multiple systems, you can experiment with programs for the motion of n attracting bodies.

The usual classification of binaries depends on the manner by which they are primarily observed:

- The members of a *visual binary* can be optically resolved. Orbital motion of one (the fainter) about the other is traced. Occasionally the proper motion of the system can be followed, when the separate orbits about the center of mass of the system can be traced.

- A *spectroscopic binary* is revealed by the oscillations of the spectral lines of the brighter of the stars, due to its orbital motion about the center of mass of the system. A line is shifted due to the Doppler effect by an amount that depends on the component of the orbital velocity in the line of sight. This radial velocity is followed as a function of the time, yielding a *velocity curve*.

- An *eclipsing binary* is revealed by changes in the magnitude (i.e., the observed brightness) of the system due to eclipses and occultations of the stars. This produces a *light curve*.

The division into three classes is somewhat arbitrary; any system, for instance, can be studied spectroscopically. An eclipsing binary is really a spectroscopic binary, oriented by fortunate accident, so that we can observe eclipses.

The first four programs in this chapter deal with the three types of binary, showing, in animation, relations between their physical and orbital parameters and the observations.

The fifth program attempts to deal with the orbital evolution of a binary when one star raises tides on the other, resulting in tidal friction. This is considered over short and long time intervals.

Finally, we consider the physical factors that go into the formation of an accretion disk in a binary system. Accretion disks are proving to be ubiquitous, and understanding how such a flat structure can be formed in a three-dimensional (3-D) environment is important. In order to do so, it is essential to learn some of the properties of the restricted problem of three bodies, and the surfaces of zero velocity (or *Roche surfaces* as they become in this stellar context). That is the intention of the sixth program. The seventh program contains a demonstration of the formation of an accretion disk.

For details about binary stars see Aitken,[1] Batten,[2] and Binnendijk.[3]

2.2 *Visual Binaries: Proper Motion*

The motion of the center of mass of a binary system is linear, with constant speed. At any instant the center of mass divides a line joining the components in a constant ratio: the ratio of the masses. You have probably seen a textbook illustration of the proper motion of a visual binary (very likely Sirius A and B). But it is another matter actually to see such motion taking place. This is what the program shows. You may select the mass ratio, the orbital elements, and the speed of the center of mass across the screen. At any instant you can cause the animation to pause, and see the line joining the stars passing through the center of mass.

Note that the phrase "mass ratio" is used consistently in all of these programs to denote the ratio of the mass of one star to the total mass of the system. This follows the long-established convention of the restricted problem of three bodies. Thus, the mass ratio conveniently lies numerically between zero and one.

Within the program the units are normalized so that the sum of the two masses is equal to **one** unit of mass, and the semimajor axis of the relative orbit is **one** unit of distance. The constant of gravitation is taken to be equal to **one**. The period of the orbit is, therefore, 2π units of the program's time. Thus, the semimajor axis is not an input quantity.

2.2.1 Running the Program

The menu items involved in running the program are

Help   **Data**   **Plot**

Help includes general information, definitions of the orbital parameters that are to be entered, and a 3-D demonstration of these parameters and orbital motion.

Data brings up the input screen, which requests the following data:

1. The mass ratio. This is the ratio of the mass of either star divided by the total mass. An input value of 0.5 would correspond to equal masses.

2. The eccentricity should be a positive number less than one. No provision is made in the program for parabolic or hyperbolic orbits since these would not be gravitationally bound.

3. The reference line for measuring the angular elements is the horizontal x-axis on the screen. The longitude of the node is measured (in this program) counterclockwise from this line. Conventionally, the longitude of the node is taken between 0° and 180°, since, without spectroscopic observation, no distinction can be made between ascending and descending motion relative to the plane of the sky.

4. The inclination should lie between 0° and 90°.

5. The argument of periastron is measured from the node in the direction of the orbital motion.

6. The speed across the screen should be positive. (Zero is acceptable.) Scaling has been arranged so that the horizontal distance across the screen is ten units of length. Since the units have been normalized, no physical interpretation can be given to this speed.

Plot provides two choices:

See Animated Proper Motion causes the paths of the stars and the center of mass to be animated. Using the F2 hot key, you can temporarily stop the motion and see a line joining the stars, passing through their center of mass.

See Animated Relative Motion enables you to see the motion of B relative to A, where A is fixed. This serves as preparation for the following program.

2.3 The True Orbit of a Visual Binary

2.3.1 The Orbital Elements of a Visual Binary

The members of a visual binary are designated primary and secondary, or A and B, A being brighter. An observation records the separation, AB, in seconds of arc, and the position angle of B relative to A. This angle is measured from the north, clockwise. See Fig. 2.1. (Note that in some texts the figure is shown inverted, with the configuration as it would appear in a telescope; in that case the angle would be measured counterclockwise.)

The observed path of B is the projection of the true orbit in the "plane of the sky." This is perpendicular to the line of sight. In Figure 2.2, E points toward the observer and N points north. With the primary at the origin and the secondary at B, the recorded position angle will be the angle between AN and AB. The

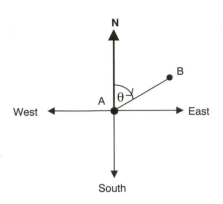

Figure 2.1: Position angle, θ.

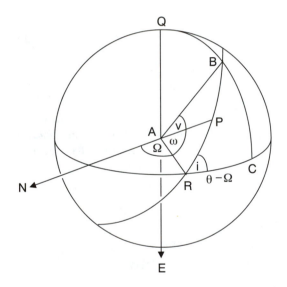

Figure 2.2: The plane of the true orbit and the plane of the sky.

projection of a conic is another conic, so the apparent orbit will be an ellipse. Ratios are not affected by projection, so the law of areas with respect to the area swept out by the line AB will be obeyed. But A will not be at a focus of the apparent orbit.
The following are the conventional orbital elements of a visual binary:

Ω The line of nodes is the intersection of the plane of the true orbit with the plane of the sky. Unless the observations are supplemented by radial velocities, the concept of an ascending or a descending node has no meaning; the nodal angle Ω is the position angle of the nodal point that lies between 0° and 180°. In Figure 2.2 it is the angle between AN and AR.

i is the inclination of the orbital plane to the plane of the sky. In the absence of radial velocities, it can be taken to lie between 0° and 90°. This will be adequate for the present program; this convention is not universal.

ω is the argument of periastron. It is measured in the plane of the true orbit from the nodal line AR to periastron, AP, in the direction of motion. It can lie between 0° and 360°. In the program for this section it will be assumed that the position angle increases with the motion.

e is the eccentricity of the true orbit.

a is the semimajor axis of the true orbit. It is measured in seconds of arc. It can be converted to astronomical units if the parallax of the system is known.

P is the orbital period in years.

T is a time of periastron passage.

2.3.2   Method of Computation

The program makes use of *Kowalsky's method*, which dates from 1873. A derivation can be found in linebreak, Smart.[12] Here we shall summarize the formulas used in the program. This method starts with the known geometry of the apparent orbit (as opposed to more practially oriented methods that deal directly with observations).

The reference system is shown in Figure 2.3. Note the disconcerting fact that the axes are *left-handed*. The equation of the apparent ellipse is

$$AX^2 + 2HXY + BY^2 + 2GX + 2FY + 1 = 0. \tag{2.1}$$

This is assumed to be known. The center, C, of the ellipse can be found. Then the line CAP is the projection of the major axis of the true orbit and P is the projection of periastron. Since ratios are not changed by projection,

$$\frac{\text{CA}}{\text{CP}} = e. \tag{2.2}$$

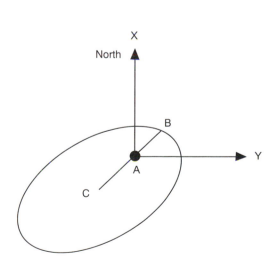

Figure 2.3: The apparent orbit of a visual binary.

The following formulas are used in the computation:

$$\frac{\cos 2\Omega \tan^2 i}{p^2} = F^2 - G^2 + A - B \tag{2.3}$$

and

$$\frac{\sin 2\Omega \tan^2 i}{p^2} = 2(H - FG), \tag{2.4}$$

where

$$p = a(1 - e^2) \tag{2.5}$$

$$\frac{2}{p^2} + \frac{\tan^2 i}{p^2} = F^2 + G^2 - A - B \tag{2.6}$$

$$\sin \omega = \frac{p}{e} \cos i(G \sin \Omega - F \cos \Omega) \tag{2.7}$$

$$\cos \omega = -\frac{p}{e}(G \cos \Omega + F \sin \Omega). \tag{2.8}$$

In the program the operator enters the eccentricity of the apparent ellipse, which is then plotted with major axis horizontal. The direction of north is, somewhat arbitrarily, taken to be vertically upward; the effect of pointing it elsewhere would be to add an angle to Ω, the position angle of the line of nodes. If x- and y-axes are defined conventionally, and the center of the apparent ellipse is at the origin, then the equation of this ellipse is

$$\frac{x^2}{a_p^2} + \frac{y^2}{b_p^2} = 1, \tag{2.9}$$

where a_p and b_p are the semimajor and minor axes. With respect to this system, let the primary star, A, be at (x_A, y_A). The ratio CA/CP is easily found, giving

$$e = \sqrt{\frac{x_A^2}{a_p^2} + \frac{y_A^2}{b_p^2}}. \tag{2.10}$$

Note that as A approaches the perimeter of the ellipse, e approaches one. It is this formula that is used in the computation of e. The coefficients in the equation of the apparent orbit with origin at A are

$$A = \frac{1}{b_p^2(e^2 - 1)}, \quad B = \frac{1}{a_p^2(e^2 - 1)}, \quad F = Ay_A, \quad G = Bx_A, \quad H = 0. \tag{2.11}$$

2.3.3 Running the Program

The menu items involved in running the program are

Datum   **Select A**   **Observe**   **Animate**

Datum shows a screen in which you are prompted to enter the eccentricity of the apparent ellipse, using a slider. This ellipse is then plotted with major axis horizontal.

Select A prompts you to pick a position for the primary by clicking on a point inside the apparent ellipse. The elements of the true orbit will then appear. The true orbit is then plotted by "unprojecting" the apparent orbit about the line of nodes through the angle i.

Observe enables you to construct a set of "observations." These might later be used in a conventional reduction procedure. The "time" is measured from the time of periastron in units of the orbital period. If more than three observations have been entered, you will be given the option of storing them in a file. (After the conclusion of the listing of this program there is listed a program READSTAR-DATA which could be used to read and print this file.)

Animate allows you to see the animated motion both in the true orbit, and, as it would be observed, in the apparent orbit. The direction of viewing this 3-D animation can be changed by the use of the **arrow** keys.

2.4 Eclipsing Binaries

This program shows the light curve of an eclipsing binary. At the same time, you will have the choice of seeing animations of the orbital motion or the eclipses. The program demonstrates the effects of changing the orbital elements of non-spherical stars and of limb-darkening. In the animations, star A is at rest.

2.4.1 Units and Parameters

Canonical units are used with the semimajor axis of the mutual orbit equal to **one** unit of length, the sum of the masses equal to **one** unit of mass, and the constant of gravitation equal to **one**. The period of revolution within the program is, therefore, 2π units of time.

Each star can be a prolate spheroid, with dimensions as input parameters. The angular rate of rotation of a non-spherical star is taken to be constant, with the axis of rotation perpendicular to the orbital plane. Realistically, non-spherical stars would only appear in conjunction with orbits of small eccentricity.

The apparent stellar magnitudes are input parameters, as are coefficients for limb-darkening. This effect is modeled in the program as follows. Let a star appear to the viewer as a projected ellipse:

$$\frac{x^2}{a^2} + \frac{y^2}{b^2} = 1. \tag{2.12}$$

At an internal point (x_i, y_i), let

$$r_i = \frac{x_i^2}{a^2} + \frac{y_i^2}{b^2}. \tag{2.13}$$

Then the luminosity is reduced by the factor

$$D\sqrt{1 - r_i^2} + 1 - D, \tag{2.14}$$

where D is the appropriate parameter: it should lie between zero and one. This "law" of limb-darkening is easily changed by editing the **Function LimbDarkening**.

The orbital parameters needed are the eccentricity, argument of periastron, and the inclination of the orbital plane to the plane of the sky. This last quantity should be close to 90° for eclipses to take place.

2.4.2 Method of Computation

The program computes, at any time, the visible area of each star. If an eclipse is taking place, it tests to find out whether this is total; if this is not so, it uses numerical integration to find the affected area. When there is no limb-darkening, a single integration suffices; but in the presence of limb-darkening, a double integral must be calculated, and, consequently, the execution of the program slows down considerably. When there is limb-darkening the "area" removed by the eclipse is weighted to allow for that effect.

Since the programming is elaborate, some explanation of the methods used will be given here. Relative to star A, let star B have coordinates (x, y, z), with the z-axis pointing away from the observer. Let the visible elliptic shapes of A and B have axes a_1, b_1, a_2, and b_2. On the screen, the center of star B will have coordinates (x, y). An eclipse is possible if

$$|x| < a_1 + a_2.$$

In this event, the sign of z determines which star is being eclipsed; the affected area is then computed in the **Procedure Eclipse**.

Figure 2.4 shows the circumstances for choosing the limits of integration in the x-coordinate. Figure 2.5 illustrates the determination of the limits of an eclipsed area in the y-direction. We shall follow the logic in the procedure. For a given value of x, let the values of y on the stellar limbs be ay1, ay2, by1, and by2, as shown.

For an eclipse to be possible, by2 < ay1.

Case a: by2 > ay2 and by1 > ay1.

Case b: by2 > ay2 and by1 < ay1.

Case c: by2 < ay2 and by1 > ay1.

Case d: by2 < ay2 and by1 < ay1 and by1 > ay2.

The *magnitude* of a star is related to the logarithm of its luminosity. If two stars have respective magnitudes M_1, M_2 and L_1, L_2, then $\log_1 0(L_2/L_1) = 0.4(M_1 - M_2)$. The magnitude of each star is entered by the operator. The program calculates luminosity using $L = 10^{-0.4M}$, and then finds luminosity per unit surface area. At any instant, the luminosity of the total observed areas is found, and the observed magnitude is calculated from this.

2.4.3 Suggested Changes to the Program

1. The orbits in the program are strictly Keplerian. In practice, some of the elements, in particular the argument of periastron, may have secular variation. This could be introduced into the program through the introduction of a variable for this variation: SecArgPCenter, say. Then in the main program

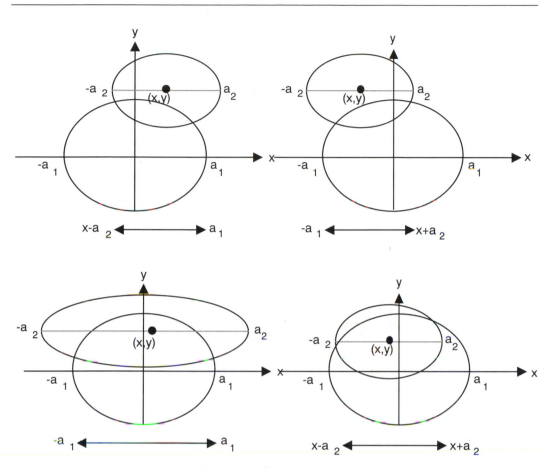

Figure 2.4: Outer limits for eclipsed areas.

for the **Procedure MakeScreen**, the angle th (which is the orientation of the major axis relative to the viewer) can be calculated from

$$th = \text{ArgPCenter} + t*(1 + \text{SecArgPCenter}),$$

2. See if you can modify the program to allow for reflection. This could be accomplished by assuming that the surface of the heated star could be separated into equal parts, each having uniform brightness. Suppose that star B is heated; the angle th in the program is measured in the direction of motion and is zero when star B appears to the right of star A (or when the z- coefficient is zero). So it passes through $\pi/2$ when star B is closest to the observer. Let the star be spherical and have radius R and let the surfaces have luminosities L_1 and L_2, L_1 being greater. If B is in front, the total observed luminosity will be

$$L_1 \frac{\pi}{2} R^2 (1 - |\cos \text{th}|) + L_2 \frac{\pi}{2} R^2 (1 + |\cos \text{th}|).$$

When it is behind,

$$L_1 \frac{\pi}{2} R^2 (1 + |\cos \text{th}|) + L_2 \frac{\pi}{2} R^2 (1 - |\cos \text{th}|).$$

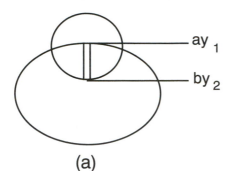

(a)

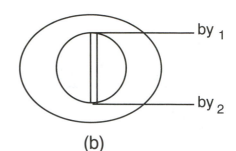

(b)

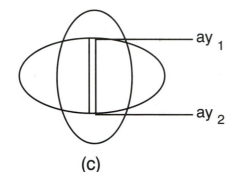

(c)

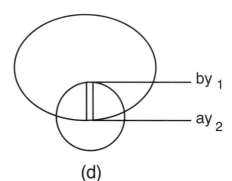

(d)

Figure 2.5: y-bounds for eclipsed areas.

2.4.4   Running the Program

The menu items involved in running the program are

Data   **Orbit**   **Eclipses**

Data shows the input screens in which you are prompted to enter the dimensions of each star together with the magnitudes and limb-darkening coefficients. The semiminor axes should not exceed the semimajor axes, and the limb-darkening coefficients should lie between zero and one. In deciding on the dimensions of the stars, bear in mind that one unit of length is the semimajor axis of the relative orbit. Certainly, the sum of the two semimajor axes should be less than one.

You are also prompted to enter the eccentricity, argument of periastron, and the inclination of the relative orbit. Remember that the inclination is the angle of the orbital plane to the plane of the sky; it should be near 90°. **Orbit** and **Eclipses** allow you to choose to see either the animated orbit or the eclipses along with the light curve.

2.5   *Spectroscopic Binaries*

2.5.1   The Velocity Curve

The elements of a spectroscopic binary are similar to those of a visual binary with the exception of the longitude of the nodes, since the line of nodes cannot be

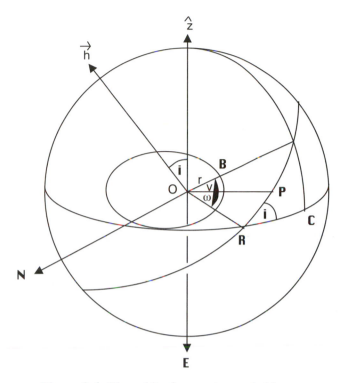

Figure 2.6: The orbit of a spectroscopic binary.

determined. The period is measured in days. Usually only one spectrum is observable; the motion of the star is followed relative to the center of mass of the system, which is at the origin, O, in Figure 2.6. OR is the line of nodes. In a coordinate system with the x-axis along OR and the z-axis parallel to $\vec{h}$ (which is perpendicular to the plane of the orbit), the star, at B, will have coordinates ($r\cos(\omega + v)$, $r\sin(\omega + v)$, 0). If this system is rotated about the x-axis through $-i$, then the new x-y plane is the plane of the sky, and the coordinates of B in this system are

$$\begin{bmatrix} 1 & 0 & 0 \\ 0 & \cos i & -\sin i \\ 0 & \sin i & \cos i \end{bmatrix} \begin{bmatrix} r\cos(\omega + v) \\ r\sin(\omega + v) \\ 0 \end{bmatrix} = \begin{bmatrix} r\cos(\omega + v) \\ r\sin(\omega + v)\cos i \\ r\sin(\omega + v)\sin i \end{bmatrix},$$

where the z-coordinate is $z = r\sin(\omega + v)\sin i$. If V_0 is the constant velocity of the center of mass in the z-direction, then the observed radial velocity is

$$V = V_0 + \frac{dz}{dt}$$

$$= V_0 + \frac{dr}{dt}\sin(\omega + v)\sin i + r\frac{dv}{dt}\cos(\omega + v)\sin i. \qquad (2.15)$$

From the formulas for elliptic motion (see Danby,[6] chapter 4) we can derive

$$\frac{dr}{dt} = \frac{nae\sin v}{\sqrt{1 - e^2}} \quad \text{and} \quad r\frac{dv}{dt} = \frac{na(1 + e\cos v)}{\sqrt{1 - e^2}}.$$

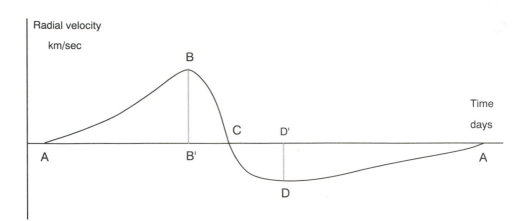

Figure 2.7: The velocity curve of a spectroscopic binary.

Substituting and simplifying, we find

$$V = V_0 + \frac{na \sin i}{\sqrt{1 - e^2}}(\cos(\omega + v) + e \cos \omega). \qquad (2.16)$$

n is the mean daily motion and is equal to $2\pi/P$, P being the period in days.

Figure 2.7 shows a possible velocity curve. The radial velocities are plotted relative to the center of mass; this assumes V_0 to be known. In principle, the determination of V_0 is straightforward, since, over one revolution, the areas above and below the line $V = V_0$ are equal. To see this, suppose that V is measured relative to V_0. Then the total area "under the radial velocity curve" is

$$\int \frac{dz}{dt} dt = z_2 - z_1$$

integrated over a complete cycle; then, due to the periodicity of the motion, $z_2 = z_1$, so the algebraic sum of the areas is zero, meaning that the areas above and below the axis are equal. In the program the velocity curve is plotted relative to the center of mass of the system.

Although the line of nodes cannot be located, the passages through the nodes are known. In the figure, B is at the maximum point on the curve, where $\omega + v = 0$. So B corresponds to the nodal passage with the star having positive radial velocity: this is the "ascending node." Similarly, D corresponds to passage through the "descending node." Further, it can be noted that the areas BB'C and CD'D are equal to the value of z at C, and the areas AB'B and AD'D are equal to the value of z at A.

2.5.2 The Mass Function

The quantities a and i occur only in the combination $a \sin i$; so while the latter may be found, a and i cannot be determined separately. a here is the semimajor axis of the orbit of the observed star about the center of mass. Let the observed star have mass m_1, and the other star have mass m_2, expressed in solar masses. The semimajor axis of the orbit of one star relative to the other will be

$$A = a(m_1 + m_2)/m_2. \qquad (2.17)$$

Now

$$(m_1 + m_2) = \frac{4\pi^2}{k^2} \frac{A^3}{P^2}, \tag{2.18}$$

where k is the Gaussian constant of gravitation. (That is, k^2 corresponds to the Newtonian constant, if units are the solar mass, the astronomical unit, and the ephemeris day.) So we can write

$$\frac{m_2^3 \sin^3 i}{(m_1 + m_2)^2} = \frac{4\pi^2}{k^2} (a \sin i)^3. \tag{2.19}$$

(This formula may appear in some references with distances converted to kilometers.) The quantity on the left is called the *mass function*. If only one spectrum is observed, this provides all the numerical information available about the masses in the system. Suppose that value of m_1 can be inferred from other data. Then it is simple to show that, if $\sin i$ is treated as a variable, m_2 will have its minimum value when $\sin i = 1$; so a lower limit can be placed on the value of m_2.

2.5.3 Suggested Changes to the Program

1. Introduce secular changes in the argument of periastron, as in the previous section.

2. Modify the program so that velocity curves for both stars are plotted.

3. See if you can use the structure of the program to construct a program for the simulation of the observation of a pulsar that is a member of a binary system.

2.5.4 Running the Program

The input units are the astronomical unit and the solar mass. The velocity curve is the radial velocity in *km/sec* and is plotted against the time in days (or hours). The menu items involved in running the program are

Data   **Orbits**

Data prompts you to enter values for the masses of the stars, the semimajor axis of the **relative** orbit, and the eccentricity, argument of periastron and inclination of the orbit. Distances are in astronomical units, and masses are in solar masses. **Orbits** then shows the orbital motion (with an arrow indicating the line of sight), an animated shifting spectral line, and the velocity curve. The velocity is measured in km/sec.

2.6 *The Tidal History of a Binary*

2.6.1 The Equations of Motion

There are many examples in astronomy where tides have had a decisive influence on the evolution of a dynamical system. In the solar system the Sun-Mercury

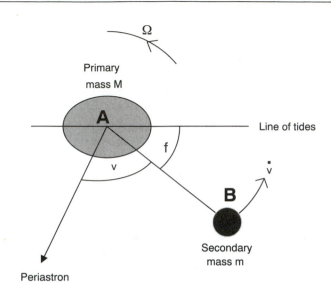

Figure 2.8: Tidal distortion. v is the true anomaly. $\Omega > \dot{v}$, so the tidal bulge is leading.

system or the Earth-Moon system are examples. Here we shall follow a simple model for investigating the effects of tides on the orbital evolution of a binary. We shall follow much of the content and notation of the paper by P. Hut.[7] This paper is a source of additional references. See Kopal[8] for detailed discussions of the dynamics of binary systems.

In the model the secondary star remains spherical, but raises tides on the primary. This is realistic if the secondary is a compact object such as a white dwarf. We shall consider the "weak friction" model, introduced by Darwin in 1879. Here the tidally distorted star takes the shape of the equipotential surface that it would have assumed a constant time, τ, earlier. So the tidal bulge **leads** the line joining the stars if the rate of rotation of the primary, Ω, exceeds the mean orbital angular velocity of the secondary, and **lags** otherwise. In Figure 2.8 the bulge is leading. (This would also apply to the Earth-Moon system.)

The lag angle as defined in Figure 2.8 is

$$f = (\Omega - \dot{v})\tau, \tag{2.20}$$

which is positive when the bulge is leading.

The potential of the tidally distorted primary will depend on many factors, including internal structure. We shall only consider the quadrupole deviation from sphericity, when it is sufficient to represent the primary by three point masses: two, with mass μ, are on the surface of the star, distant R from its center; and the third, with mass $M - 2\mu$, is at the center, as shown in Figure 2.9.

μ is proportional to the magnitude of the tides, which is proportional to m, the mass of the secondary, and inversely proportional to the cube of r, the distance between the stars. Hut takes

$$\mu = \frac{1}{2}mk\left(\frac{R}{r}\right)^3. \tag{2.21}$$

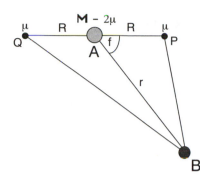

Figure 2.9: Gravitational approximation for the primary.

k is a parameter that depends on the degree of concentration of the primary; higher concentration corresponds to lower k. Realistically, $k \sim 0.1 - 0.01$. Because of tidal dissipation, r must be evaluated, not at the current time t, but at time $t - \tau$. So we have

$$\mu(t) = \frac{1}{2} km \left(\frac{R}{r(t - \tau)} \right)^3 . \tag{2.22}$$

Assuming that τ is not too great, it is adequate to approximate $r(t - \tau)$ as $r(t) - \tau \dot{r}(t)$. This approximation will be used in the calculations.

Let $\hat{\rho}$ be the unit vector parallel to $\overrightarrow{AP}$. Then, if $\overrightarrow{AB} = \vec{r}$, the equation of motion of the secondary is

$$\frac{Mm}{M + m} \ddot{\vec{r}} = -G \frac{(M - 2\mu)m}{r^3} \vec{r} - G \frac{\mu m}{\|\vec{r} - R\hat{\rho}\|^3} (\vec{r} - R\hat{\rho})$$

$$- G \frac{\mu m}{\|\vec{r} + R\hat{\rho}\|^3} (\vec{r} + R\hat{\rho}) . \tag{2.23}$$

Here $Mm/(M + m)$ is the reduced mass.

We shall assume that the orbital inclination of the secondary is small, so that the axis of rotation of the primary does not change. We shall use a coordinate system with the z-axis parallel to this axis of rotation. The orbital angular momentum of the system resolved in the z-direction is

$$H_z = \frac{Mm}{M + m} (x\dot{y} - y\dot{x}) . \tag{2.24}$$

Let the moment of inertial of the primary be $I = MR^2 r_g^2$. r_g depends on the concentration of the star; for a homogeneous star, $r_g^2 = 2/5$. For a centrally condensed star, $r_g^2 \sim 0.1$–0.01. The rotational angular momentum of the primary is $I\Omega$. Since the total angular momentum of the system is conserved, we must have

$$\frac{d}{dt} (H_z + MI\Omega) = 0 . \tag{2.25}$$

Hence,

$$\dot{\Omega} = \frac{m}{M + m} (y\ddot{x} - x\ddot{y})(R^2 r_g^2)^{-2}. \tag{2.26}$$

Eq. 2.23 and Eq. 2.25 define the model.

2.6.2   The Averaged Equations

In the program the equations of the model are solved numerically, and at each step the osculating Keplerian elements are computed. Many integration steps are required for each revolution, and the visual appearance of the graphed elements contains the orbital short-term wiggles as well as the longer-term trends. In these trends, the changes of nodes and line of apsides are relatively fast compared with those of the remaining elements.

A method often used under these circumstances is to average the right-hand sides of the equations for da/dt, de/dt, di/dt, and $d\Omega/dt$. analytically over one revolution. The short-term effects are then smoothed out. Much larger integration steps can be used. Hut has carried out this procedure. In running the program, one option is to solve his averaged equations.

2.6.3   Suggested Change to the Program

In his paper Hut carries out an additional averaging, leading to equations for the slow variations of the semimajor axis and eccentricity. Model the Procedure See-AveragedElements to solve these equations. You will have Neq = 2, a new Procedure Fun, and the choice of only two elements to be plotted. Confirm Hut's results and see if you can extend them.

2.6.4   Running the Program

Two of the parameters, k and r_g, depend on the degree of concentration of the primary. The value of k is important in controlling the rate at which results can be observed. The parameters have been set in the program with the values $k = 0.1$ and $r_g = 0.01$.

When entering parameters, bear in mind the approximations that have been made in deriving the equations for the model. A wide range of projects is possible; but it is easy to produce results that are nonsensical.

The menu items concerned with runing the program are:

Data   **See Elements**   **Elements**

In the first input screen you are prompted for the mass or the primary, in solar masses, and the mass and orbital elements of the secondary.

In the second input screen you are prompted for the radius of the primary, in units of the solar radius; the value of τ; and the initial angular velocity of rotation of the primary in radians per day. You will be shown the value of the mean orbit motion of the secondary in the same unit, to help you pick suitable values.

See Elements gives you two choices: **Not Averaged. See Each Revolution** and **Averaged**. These correspond to the options already described. When you make a choice, you will be prompted for the length of time over which you want to take the numerical integration. You will be shown the period of the binary orbit, to help pick a suitable value. While the numerical integration is taking place, the osculating eccentricity will be plotted.

Elements lists the elements that you can see, appropriate to the option you have chosen. Note that if the inclination is zero, neither the longitude of the ascending node nor the inclination can be seen. You also have the option of extending the time, to do a further integration.

2.7   The Restricted Problem of Three Bodies

The model to be described here is, strictly, the "circular" restricted problem of three bodies. Two bodies revolve around each other in circular Keplerian orbits, while a third, having mass too small to infuence the first two, moves in their combined gravitational field. This model approximates several instances in dynamical astronomy, and will be applied here to the transfer of matter between members of close binaries. Accordingly, the two masses will be called "stars."

2.7.1   Units

A canonical set of units is almost invariably used in formulating this problem:
   The sum of the masses of the stars is **one** unit of mass. Then the separate masses are usually written as μ and $(1 - \mu)$. Conventionally, $\mu \leq 0.5$.
   The constant distance between the stars is **one** unit of length.
   The period of the circular orbits is 2π units of time. This is equivalent to taking the constant of gravitation to be equal to **one**.

2.7.2   Equations of Motion

The equations are formulated with respect to a rotating reference system in which the stars remain fixed. With the origin at the center of mass, the coordinates of the stars with masses μ and $(1 - \mu)$ are $(-\mu, 0, 0)$ and $(1 - \mu, 0, 0)$, respectively. Let the particle, at P, have coordinates (x, y, z). Then it experiences the gravitational force

$$\nabla U = \nabla\left(\frac{1 - \mu}{\rho_1} + \frac{\mu}{\rho_2}\right), \tag{2.27}$$

where

$$\rho_1 = \sqrt{(x + \mu)^2 + y^2 + z^2} \quad \text{and} \quad \rho_2 = \sqrt{(x - 1 + \mu)^2 + y^2 + z^2}. \tag{2.28}$$

The reference system is rotating with unit angular speed; so the equations of motion are

$$\left.\begin{array}{l} \dfrac{d^2x}{dt^2} - 2\dfrac{dy}{dt} - x = \dfrac{\partial U}{\partial x} \\[2mm] \dfrac{d^2y}{dt^2} + 2\dfrac{dx}{dt} - y = \dfrac{\partial U}{\partial y} \\[2mm] \dfrac{d^2z}{dt^2} \qquad\qquad = \dfrac{\partial U}{\partial z} \end{array}\right\}. \tag{2.29}$$

Note the Coriolis and centrifugal terms on the left.

2.7.3   Jacobi's Integral and Surfaces of Zero Velocity

Multipying the members of Eq. 2.29 by dx/dt, dy/dt, and dz/dt, respectively, and adding, we find

$$\frac{d^2x}{dt^2}\frac{dx}{dt} + \frac{d^2y}{dt^2}\frac{dy}{dt} + \frac{d^2z}{dt^2}\frac{dz}{dt} - x\frac{dx}{dt} - y\frac{dy}{dt} = \frac{\partial U}{\partial x}\frac{dx}{dt} + \frac{\partial U}{\partial y}\frac{dy}{dt} + \frac{\partial U}{\partial z}\frac{dz}{dt} = \frac{dU}{dt}.$$

(2.30)

Integrating,

$$\frac{1}{2}\left[\left(\frac{dx}{dt}\right)^2 + \left(\frac{dy}{dt}\right)^2 + \left(\frac{dz}{dt}\right)^2\right] - \frac{1}{2}(x^2 + y^2) = U + \text{constant}. \quad (2.31)$$

Define

$$\Phi(x, y, z) = \frac{1}{2}(x^2 + y^2) + U + \frac{1}{2}\mu(1 - \mu). \quad (2.32)$$

Equation 2.31 can then be written as

$$\left(\frac{dx}{dt}\right)^2 + \left(\frac{dy}{dt}\right)^2 + \left(\frac{dz}{dt}\right)^2 = 2\Phi - C. \quad (2.33)$$

This is **Jacobi's integral**.

For a given value of C, the expression $2\Phi(x, y, z) - C$ may be positive or negative, depending on the position (x, y, z). For motion to be possible, the expression must be non-negative. Therefore the surface

$$2\Phi(x, y, z) - C = 0 \quad (2.34)$$

will, if it exists, separate regions in space where motion is possible from regions where it cannot take place. Jacobi's integral is known as an *isolating integral* for this reason.

Figure 2.10 shows sections of these surfaces in the x-y plane for three different values of C. The outer curve is the section of what is, topologically, a cylinder; the inner figure-eight, which shares the same value of C with the previous one, is the section of two spherical regions. Surfaces within these (having higher values of C, or lower energy) are topologically similar to spheres, so that if the particle is within one of these, it is trapped.

These surfaces are important when considering the possibility of matter passing from one star in a binary system to the other. Such a passage can lead to the formation of an accretion disk. In the context of binary stars the surfaces are referred to as Roche surfaces, and, indeed, the restricted problem of three bodies is sometimes referred to as the Roche model.

The singular points in the diagram correspond to points where the time derivatives in Eq. 2.29 are zero, and so are equilibrium solutions, known as *Lagrangian points*. They are labelled L_i, $i = 1 \ldots 5$. For given μ, each will have a corresponding value, C_i, of C. As C decreases, the energy increases. For low energy, with $C > C_1$, the particle is trapped insider a sphere around one of the stars, or must lie outside the outer cylinder. At $C = C_1$ the figure-eight is formed, with the crossing point at L_1. For $C_1 < C < C_2$, the particle is free to move between the masses (or must lie outside a cylinder). At $C = C_2$ this inner, closed region and the outer

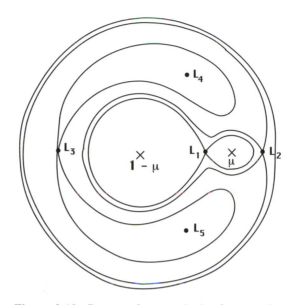

Figure 2.10: Curves of zero velocity for $\mu = 0.1$.

cylinder intersect at L_2. For higher energies, an opening appears around L_2 so the particle may escape from the vicinity of the stars. Further shrinking of the forbidden region occurs until it splits into two, at the point L_3, and finally disappears at the triangular points, L_4 and L_5. The value of C at these points is $C_4 = C_5 = 3$, by virtue of the definition, Eq. 2.32, of Φ. See Danby[6] for more details.

2.7.4 Stability of the Lagrangian Points

In the program you will have the option to see orbits, and to verify that they are subject to the limitations given by the curves of zero velocity. You will also be able to observe orbits starting with low velocity close to the Lagrangian points. Understanding such motion close to L_1 is indispensable for understanding the formation of an accretion disk.

The three equilibria L_1, L_2, L_3 are unstable; orbits leaving them, with low relative speed, do so in disciplined streams. The triangular points are stable for sufficiently small μ. Specifically, they are stable for $\mu < \mu_1 = 0.03852\ldots$ with the exception of the two values $\mu_2 = 0.02429\ldots$ and $\mu_3 = 0.01352\ldots$.

2.7.5 The Photo-Gravitational Restricted Problem of Three Bodies

If the massive bodies are stars and the third body is small in size, it will be subject to pressure from the stellar radiation. This can be modeled by modifying Eq. 2.27 so that

$$U = \frac{\alpha(1 - \mu)}{\rho_1} + \frac{\beta\mu}{\rho_2}, \qquad (2.35)$$

where α, $\beta \leq 1$. In the program α and β must be positive; this ensures that the three collinear Lagrangian points continue to exist. You should confirm that for the triangular points

$$\rho_1 = \alpha^{1/3} \quad \text{and} \quad \rho_2 = \beta^{1/3}. \tag{2.36}$$

For these to exist, it is necessary that $\alpha^{1/3} + \beta^{1/3} < 1$. So with increasing radiation pressure these points approach the y-axis, eventually coalescing at L_1. A point of interest is that provided that both stars exert radiation pressure, it is possible for L_1 to be stable. For further details, and a full discussion of the stability of these points, see the paper by J. F. L. Simmons et al.,[11] which is also a source of additional references.

We take the modified version of Jacobi's integral, Eq. 2.33, to be

$$\left(\frac{dx}{dt}\right)^2 + \left(\frac{dy}{dt}\right)^2 + \left(\frac{dz}{dt}\right)^2 = x^2 + y^2 + 2\frac{\alpha(1-\mu)}{\rho_1} + 2\frac{\beta\mu}{\rho_2} - C + \mu(1-\mu)$$
$$+ 3(1 - \alpha^{2/3}(1-\mu) - \beta^{2/3}\mu). \tag{2.37}$$

Then the value of C will be equal to 3 when Equations 2.36 are satisfied.

A feature of this model, as you will find out, is that the figure-eight zero velocity curve may cease to exist, and the formation of an accretion disk, by the mechanism descibed in the following section, may not be possible.

2.7.6 Poincaré Maps

A hundred years ago Henri Poincaré[9] devised a method of investigating dynamical systems such as the 2-D restricted problem. The method, described for some time as the method of "surface of section" received considerable theoretical application. With the advent of high-speed computing it then assumed a major role in the numerical investigation of theoretical aspects of dynamical systems.

The method is applied to conservative dynamical systems having two degrees of freedom, of which the planar, circular restricted problem is an example. The phase space of such a system is four-dimensional. The system will have an integral, like Jacobi's integral of the form

$$F(x, y, v_x, v_y) = C. \tag{2.38}$$

Suppose that a value for C is chosen, and kept fixed during an experiment. Then if any three of the phase variables are specified, the fourth can be found. Effectively, then, the phase space has been reduced to three dimensions.

To follow an orbit in these three dimensions, Poincaré defined a two-dimensional surface and followed successive crossings of the orbit across that surface. To be more specific, let the three dimensions chosen be (x, y, v_x). Let the surface be the plane $y = 0$. Let an orbit cross this surface with $v_y > 0$ at a point P_1. The next crossing will have $v_y < 0$ and is (for this discussion) ignored. Let the following crossing, again with $v_y > 0$, be at the point P_2. Then P_2 is called the "Poincaré map" of P_1.

Poincaré introduced this construction in order to discuss what we now call "chaotic systems." He accurately predicted many properties of such systems. A non-chaotic system will have a second integral of motion, and, in general, the orbit in the three-dimensional space will lie on a torus. The plane of section will intersect this torus in a closed curve, called an *invariant curve*; so successive Poincaré mappings will produce points lying on such a curve. In a computation, such behavior is fairly easy to recognize. If the successive points do not lie on any closed curve (or maybe a set of these), the system is acting in a chaotic manner.

It has long been known that a system of the sort under discussion will possess only one integral (that is, a function of x, y, v_x, v_y, and possibly the time that is constant for **all** solutions). In fact it has been proved that, in a certain sense, the set of integrable systems has measure zero. The restricted problem is a chaotic system. But many solutions appear to behave in a non-chaotic way. Mathematically it may be shown that invariant may curves exist, but that between these curves the mappings are chaotic. If the invariant curves are close to one another, we shall not see any signs of chaos unless we use "high magnification" numerical tools.

Typically, we investigate a system by changing a parameter; it might, in this instance be C or μ. For a range of values of the parameter, the solutions look non-chaotic. Then the invariant curves break up and chaos takes over.

In this program you will have the option of constructing these maps and observing the properties just mentioned.

2.7.7 Suggested Changes to the Program

1. As programmed, the Poincaré maps find crossings for which $y = 0$. This may be too restrictive. For instance, if motion close to a triangular Lagrangian point is to be investigated, then $y \neq 0$. Suppose that you are to look for crossings at $y =$ yCross. Make the following changes:
   In the Procedure StartPoincareMap,

$$\textbf{y0 : = 0; becomes y0 : = yCross;}$$

   In the Procedure FindPoincareCrossing,

$$\textbf{StepSize : = } - \textbf{LocalState[3]/LocalState[4];}$$

becomes

$$\textbf{StepSize : = } -\textbf{(LocalState[3]} - \textbf{yCross)/LocalState[4];}$$

with repetition until

$$\textbf{(Abs(LocalState[3]} - \textbf{yCross)} < \textbf{0.0001) OR (Counter = 10);}$$

In the Procedure PlotPoincare, **State[3]** must be initialized to **yCross**. The condition for a crossing to have taken places is then

$$\textbf{If CrossSign*(State[3]} - \textbf{yCross)} < \textbf{0 THEN. . .}$$

For motion close to L4, take yCross $= \sqrt{3}/2$. Note that $L4$ is unstable for

$$\mu > \mu_1 = \frac{1}{2}(1 - \sqrt{69}/9) \simeq 0.03852 \tag{2.39}$$

and also for the special values

$$\mu = \mu_2 = \frac{1}{2}(1 - \sqrt{611}/15\sqrt{3})) \simeq 0.02429 \tag{2.40}$$

and

$$\mu = \mu_3 = \frac{1}{2}(1 - \sqrt{213}/15)) \simeq 0.01352 . \qquad (2.41)$$

The Jacobi constant, C, should be taken to be very close to 3. The Poincaré maps can provide considerable insight into the mechanics as stability and instability are interchanged.

2. In following the Poincaré maps, *every* crossing, with $y = $ yCross is recorded. It may be clearer if you only record those for which the derivative of y, i.e., State[4], has the same sign. Then a point in the diagram corresponds to one and only one orbit.

3. For extra clarity you may want to magnify the screen containing the Poincaré maps. This is ViewPort(6). Redefine it as "DefineViewPort(6, 0, 1, 0.05, 0.95);". Now eliminate reference to ViewPort(5), and, in Procedure PlotPoincare, remove the line

    PlotLine(xOld, yOld, xNew, yNew);

    Also, in Procedure StartPoincareMap, erase the call to **MouseBorders**.

4. The demonstration program in the **Help** menu might be enlarged into a program in its own right, with 3-D rotations, and a choice of parameters.

5. If resistance proportional to the square of the speed relative to an interstellar medium is included, the triangular Lagrangian points become unstable for any mass ratio. Modify the program to include such resistance to verify this result, and carry out an analytical verification.

6. The input of starting conditions using the mouse is easy but imprecise. Consider modifying the program so that input is from the keyboard.

7. Poincaré maps can be used to investigate the motion of an asteroid in resonance with Jupiter. This investigation might be made in conjunction with the program in the n-body module dealing with this topic. Jupiter's orbit must be circular for the restricted three-body problem to be used. Input for initial conditions of asteroid orbits should be from the keyboard. A circular orbit of an asteroid would appear in the Poincaré map as an isolated dot, or a "fixed point" that is mapped into itself. Orbits close to this are of interest. Parameters to be varied are different resonances and the mass of Jupiter.

8. An similar investigation concerns the possibility of stability of planetary orbits in a Sun-Jupiter system if the mass of Jupiter is increased. This could be relevant if the mass of a "proto-Jupiter" was considerably higher than the present mass of the planet.

2.7.8   Running the Program

The sections of the menu that control the running of the program are

Data   **New Energy**   **Orbit**   **CleanUp**

Data opens an input screen in which you are prompted for a value of the mass ratio, μ. Remember that this is the ratio of the mass of one star to the total mass. It should lie in the interval $0 < \mu < 1$. You can also enter a number for the spacing of the grids used in calculating the Roche curves. The default values are 50 along each axis. (The highest permissible values are 70.) Higher numbers lead to better resolution but slow down execution of the program. You will also be prompted to enter extreme coordinates for the plotting of the Roche curves. (Later, during this plotting, you will have the option of zooming in or out around selected points.)

After the input has been entered, there is a delay while the grid points are calculated. In the opening figure, positions of the five Lagrangian points are shown.

New Energy allows you either to enter a value of C and to see the appropriate Roche curve, or to click on a point and see the Roche curve that contains this point. This latter option is useful, for instance, if you want the curve close to L_1.

Orbit allows you to plot trajectories in one of two ways. You can click on a starting point and then click at a second point to define the direction of the starting velocity; the speed will be determined consistent with the current value of C. To see an orbit close to a trianglar point, enter a value of C **very** close to 3 (such as 2.99999.), and then click on a starting point near to the triangular point.; putting $C \neq 3$ will ensure that the initial speed is small.

Alternatively, you can see the orbit and the Poincaré maps. The starting point for an orbit is determined by clicking at a point in the (x, v_x) plane of the Poincaré map, for which $y = 0$. (The value of v_y is found from Jacobi's integral.)

2.8   The Formation of an Accretion Disk

2.8.1   The Model

This program is designed to illustrate the mechanics of the formation of an accretion disk in a binary system. This is the basic model for the class of objects known as cataclysmic variables (CVs). The word "cataclysmic" should be compared to "catastrophic." After a catastrophic event, such as a supernova explosion, a star is never the same again. But, while it may be violent at the time, a cataclysmic event is followed by a return to the pre-explosion state. In such an explosion about 10^{-4} of the star's mass may be lost, and the explosion may recur after an interval of some 10^4 years. This is the case with classical novae, where the outburst has been compared to a star "sneezing." This recurring process involves the transfer of matter from one star to the other. The model was proposed by Crawford and Kraft in 1956. A good description of the basic physics can be found in chapter 2 of *Classical Novae*,[4] edited by M. F. Bode and A. Evans; the article is written by A. R. King. Also see *Interacting Binary Stars*,[10] edited by J. E. Pringle and R. A. Wade.

For a nova, the primary star is a white dwarf. The period may be a few hours, so the separation may be of the order of ten times the solar radius. The stars are close enough for it to be assumed that the orbits are circular, and condensed enough for the circular restricted problem of three bodies to be adopted to describe the dynamics. In this context, this is called the Roche model. It is assumed that you have gained experience with this model through the preceding program.

The crucial Roche surface is the based on the figure-eight section through $L1$. $L1$ divides the surface into two volumes called Roche lobes. For matter to be transferred from the secondary, that star must fill its Roche lobe. This is the first stage of the process.

The second stage begins with matter leaking into the lobe of the primary. It must do so through $L1$, and will, close to $L1$ have low speed. It will then follow a well-defined path, which will lie in, or very close to, the plane of the orbital motion. So, for our purposes, the subsequent dynamics can be considered to be two-dimensional. Let us assume that the leakage has just started. Because of the small size of the white dwarf and the rotation of the system, the matter will not fall onto the primary, but will travel around it. If the only forces were gravitational, then the material would travel in a precessing path that can be followed in the preceding program. But, after one revolution around the primary, the matter will meet the continuing stream from $L1$. This is the third stage.

In talking about the motion around the white dwarf it is convenient to ignore the secondary, and to think in terms of Keplerian motion. The collision of the streams of gas will result in loss of kinetic energy to radiation. This is shown by the hot spots that are observed where the incoming stream hits the accretion disk. If energy is lost, but angular momentum preserved, then particles will tend to find the orbits having given angular momentum but minimum energy; these orbits are circular. So collisions will lead to a hot spot, the circularization of the orbits and the heating of the disk.

Overall, loss of energy will lead to orbits moving closer to the white dwarf. So the disk will expand inward. But these shrinking orbits will be losing angular momentum. This must be transferred to particles in the outer orbits, so those orbits will tend to expand, preserving the overall angular momentum. So the disk expands inward and outward, and will continue to heat up.

These arguments are general. They are backed up by some equations in the article by A. R. King, already cited. The following chapter in the same book, written by S. Starrfield[4], contains following statement: "Some viscous process, as yet unknown, acts to transfer material inwards and angular momentum outwards so that a fraction of the material lost by the secondary ultimately ends up on the white dwarf." No attempt has been made in the program to involve any physics in the expansion of the disk; it simply expands until it reaches the surface of the primary.

Possibilities that follow this state of affairs are even more general and varied, as the article by Starrfield shows. Given the right conditions, the temperatures in the gas close to the surface of the white dwarf can reach 10^8 K; this leads to what is called a "thermonuclear runaway." It can result in energies sufficient to eject material and release energy consistent with observed nova outbursts. Once more, no attempt is made in the program to represent these processes.

2.8.2 Suggested Change to the Program

Introduce radiation pressure, and investigate conditions for the continuing formation of a hot spot.

2.8.3 Running the Program

The first menu item involved in running the program is **Data**, which prompts you for the mass ratio,

$$\mu = \frac{m_2}{m_1 + m_2}.$$

(This should not be confused with the ratio m_2/m_1, which appears as q in the cited article.) In all known cases, $\mu < 0.5$, but there is nothing to prevent you from trying other values.

The second item is **Next**, which allows you to click through the sequence

Fill Lobe B,

Matter Leaks Into Lobe A,

Expansion of the Accretion Disk, and

Heating of Accretion Disk.

Please remember when running the program that its purpose is to model the production of an accretion disk, and not its subsequent history.

2.9 Exercises

Visual Binaries

2.1 **Orbital elements**

Become thoroughly familiar with the meanings of the elements of the orbit. Semimajor axis and eccentricity are the physically significant quantities; the three Eulerian angles describing the orientation of the orbit in space are accidental. All affect the observed motion. Start with a circular orbits, i.e., zero eccentricity; make a succession of runs **only** changing the eccentricity. Make notes on how the observed orbits are affected. Include in your observations the motion of B relative to A, as well as the proper motions.

Repeat the exercise, varying just one other element.

2.2 **Law of areas**

Kepler's law of areas states that in the relative orbit of two bodies the line joining the bodies sweeps out equal areas in equal intervals of time. This remains true even if we are looking at the *apparent* orbit. Try to give a convincing argument showing why this must be so.

To demonstrate this, use the option of showing the animated relative motion. Using the hot key F6, slow down the animation. Now, use the F2 hot key to stop and start the motion. Try to ensure that the intervals in which the motion is taking place are equal: a ticking clock or metronome will help. Your results should confirm the law of areas in the apparent orbit.

2.3 True and apparent orbits

Ratios are not changed by projection. Figure 2.3 shows an apparent orbit, and star A. The projection of an ellipse is another ellipse, so the apparent orbit will always be an ellipse. Explain why

a. The center, C, of the apparent ellipse must also the the center of the true ellipse.

b. The line through C and A will cut the apparent ellipse in a point P, which must be the projection of periastron in the true ellipse.

c. The ratio CA/CP is equal to the eccentricity of the true ellipse.

Confirm, by experiment, that this is true. Verify theoretically that these properties are correct.

2.4 Parallax

The output of the program can be incorporated into calculations; but first, the figure must be scaled so that "lengths" are in seconds of arc, and a period for the orbit must be prescribed.

When the results of the computation of the true orbit appear, a scale appears at the bottom of the screen; it has been chosen so that the semimajor axis of the apparent orbit is 2 units of length. However, in an astronomical application, the semimajor axis is expressed in seconds of arc: i.e., it is the angle subtended by the semimajor axis as viewed by the observer. The interpretation of this scale will be up to you. Suppose that one unit on the scale corresponds to $0''.1$; then the semimajor axis of the true orbit will be expressed in this unit. So if the value of the semimajor axis that appears on the screen is 3.5, then $a = 0''.70$. To apply this, consider three quantities: P, the orbital period in years; p, the parallax of the system, in seconds of arc; and $(m_A + m_B)$, the sum of the masses of the two stars. If a is expressed in seconds of arc, then the semimajor axis of the true orbit in astronomical units is

$$a_1 = \frac{a}{p}. \tag{2.42}$$

Then, from Kepler's third law (with the year as the unit of time)

$$(m_A + m_B)p^3P^2 = a^3. \tag{2.43}$$

Suppose the period is $P = 1$ year. If $a = 0''.70$ and we assume $(m_A + m_B) = 2$ solar masses, then the *dynamical parallax* is $p = 0''.56$. If the spectral types of A and B are known, then a better value might be assumed for $(m_A + m_B)$. If the parallax is known to be, say, $p = 0''.4$, then the sum of the masses can be calculated to be $(m_A + m_B) = 5.4$ solar masses.

You have a choice of what you can assume and what you can discover. Investigate systematically the changes in what you find while the location of start A is systematically changed. Look for diagrams of apparent orbits in the literature, and apply this program to them.

2.5 Observations

One option in the program for finding a true orbit is to manufacture "observations" of position angle and separations. These can then be used in a conventional program for the reduction of observations. (For information about such programs see references Couteau[5] and Smart[12].) To make use of these observations you will need to scale the separations, as described above. The "times" of observation that appear are fractions of the orbital period, measured from periastron passage. To make them usable, an actual period will need to be prescribed.

Eclipsing Binaries

2.6 Light Curves

Textbooks show light curves for a limited number of parameters. A prime use of this program is to investigate systematically the effects of changing the parameters. I suggest that you start with spherical stars having no limb-darkening, and investigate changes in the orbital parameters. Then one parameter at a time should be gradually changed, and its effects fully discussed. Bear in mind that, in practice, eccentricities are low for eclipsing binaries. When you have gained some experience, see if you can sketch a light curve **before** you have seen it on the screen.

2.7 Reflection

The principal effect that has been omitted from the program is "reflection"; one star may be heated by the other, so that the surface pointing toward the other star is brighter than that pointing away. Start with any light curve; how would this be modified if reflection were to be an additional factor?

2.8 Matching light curves

In textbooks and in the literature, look up light curves of observed eclipsing binaries. See if you can replicate them using the program. If this is not possible, can the deficiencies be explained by allowing for reflection?

2.9 Apsidal motion

Suppose that there is a precession of the line of apsides. How would the light curve be modified as this took place?

2.10 Accretion disk

Start with a light curve for a circular orbit. How would this be modified if one star had an accretion disk and a hot spot?

Spectroscopic Binaries

2.11 Effects of changing elements

As in the case of eclipsing binaries, systematically explore the effects of varying elements, one at a time.

2.12 Predicting velocity curves

As before, see if you can sketch a velocity curve **before** you have seen it on the screen.

2.13 Matching velocity curves

In textbooks and in the literature, look up velocity curves of spectroscopic binaries and see if you can replicate them using the program.

2.14 Apsidal motion

Suppose that there is a precession of the line of apsides. How would the velocity curve be modified as this took place?

2.15 Effects of changing a and i

It is stated above that only the combination $a \sin i$ can be determinind from the velocity curve. Investigate this assertion by varying a and i while keeping $a \sin i$ fixed.

The Tidal History of a Binary

2.16 Introduction

As a preliminary to this, as with any project, consult the literature concerning the nature of what you can expect. For a good beginning here, see the bibliography in the article by Hut.[7] Many of these references end with rather general results. See if you can replicate them through the numerical solutions.

2.17 Apsidal motion

Use the non-averaged equations to investigate the apsidal motion of close binaries. For this, start with synchronous motion, with $\Omega = n$, the orbital mean motion, and $\tau = 0$. Let the eccentricity and the inclnation be small. If the motion is nearly synchronous, but still $\tau = 0$, then results will not change qualitatively, and Ω will be constant. Now introduce a small value for τ. The orbit should become synchronous, and the eccentricity should approach zero. You will be seeing the approach of the model to an equilibrium which will have the character that it is synchronous, with eccentricity and inclination zero.

2.18 Effects of changing parameters

The best rule is to change just one parameter at a time, and to change it gradually. For instance, gradually increase the initial Ω of the primary. Energy should be transferred from the spinning star to the orbit. Then decrease the initial Ω to observe the reverse effect. Or increase the initial eccentricity or inclination. But remember that the inclination is assumed to be small.

2.19 Averaged equations

When you are satisfied that you have seen the short-term trends, change to the option in the program for seeing the averaged calculations. Repeat the general investigations.

2.20 **Angular momentum of the system**
In terms of the quantities, the total angular momentum of the system can be written as

$$L = I\Omega + \frac{Mm}{M + m}\sqrt{G(M + m)a(1 - e^2)}$$

$$= I\Omega + \frac{Mm}{M + m}\sqrt{G^{4/3}(M + m)^{4/3}\, n^{-2/3}(1 - e^2)}$$

$$= I\Omega + G^{2/3}Mm(M + m)^{-1/3}n^{-1/3}(1 - e^2)^{1/2}.$$

Here n is the orbital mean motion. We have used the formula $n^2a^3 = M(M + m)$. If an equilibrium is reached, then $e = 0$ and $\Omega = n$. Let the equilibrium value of Ω be Ω_0. Then

$$L = MR^2r_g^2\Omega_0 + G^{2/3}Mm(M + m)^{-1/3}\Omega_0^{-1/3}. \tag{2.44}$$

L can be found from the initial conditions, and this equation can then be solved for possible Ω_0. (This is best done using a hand calculator with a solver facility, or you might write a computer program.) When Ω_0 has been found, the equilibrium a_0 can be found since $\Omega_0^2 a_0^3 = M(M + m)$. As a project, investigate the properties of the solutions of Eq. 2.44.

2.21 **Stability of the system**
Introduce

$$\alpha = \frac{h_0}{I\Omega_0} = \frac{m}{M + m}\frac{A_0^2}{R^2r_g^2},$$

the ratio of the orbital and rotational angular momenta at the equilibrium. A result quoted by Hut[7] is that the equilibrium is stable if and only if $\alpha > 3$. Investigate this situation using initial conditions giving α on each side of the inequality. What happens when the orbit is unstable?

2.22 **Further averaging**
Hut carries out a further averaging, reducing the dependent variables to the semimajor axis and eccentricity. He uses these to investigate tidal evolution starting far from the equilibrium. These equations have not been included in the present program. But look at his results and see if you can replicate and extend them with the existing program.

The Restricted Problem of Three Bodies

2.23 **Curves of zero velocity**
Initial experiments should be done without radiation pressure.

In preparation for understanding a mechanism for the formation of an accretion disk you should become thoroughly familiar with the curve of zero velocity and the regions of space where motion is permitted. Use values of $\mu = 0.1, 0.2, 0.3, 0.4, 0.5$. For each value, take values of C sufficient to produce a diagram like Figure 2.10.

2.24 Lagrangian points

When working with this model it is convenient to be able to calculate the locations of the Lagrangian points and the values of C associated with them. Construct a program to do this, or use a calculator.

2.25 Matter leaking through L_1

Matter leaking from one star to another in a binary system is likely to do so through a small gap at L_1 when the figure-eight curve of zero velocity is barely open at that point. To construct such a figure-eight curve, use the **Click** option for choosing the energy and click just above L_1. Now plot an orbit by clicking first at L_1. Repeat this experiment for a range of values of μ. Interpret the "orbit" as a stream of particles. When the head of the stream hits the incoming stream, the interaction of the gases creates a "hot spot," which is an observable feature of many accretion disks.

2.26 Stability of L_4 and L_5

Although they are not relevant to accretion disks, the triangular points, L_4 and L_5 are of considerable astronomical interest. They are unstable points of equilibrium for $\mu > \mu_0 = 0.03852$, and for $\mu_1 = 0.02429$ and $\mu_2 = 0.01352$; otherwise they are stable. To see the character of orbits close to these points, when there is stability, first choose a suitable value of μ. You might start with $\mu = 0.001$, relevant to the Sun-Jupiter-Trojans configuration. Next, choose a value of C very close to 3 (3.00001, for instance) or click on a point close to L_4. Then observe orbits by first clicking close to L_4. The pattern of the stable orbits is similar to the pattern of the distribution of the Trojan asteroids around L_4 and L_5. Experiment also with $\mu = \mu_1$ and $\mu = \mu_2$, and values nearby to see if you can learn anything about the nature of the instability for these special values of μ.

2.27 Restrictions on orbital motion

When plotting trajectories you may notice that although, by virtue of Jacobi's integral, the motion might occur anywhere inside the appropriate curve of zero velocity; in fact, it seems to be confined to a smaller but well-defined region. This seems to hint at the existence of another isolating integral. A considerable amount of work has been done, especially on galactic orbits, concerning the existence and possible roles of such extra integrals; they are often referred to as "third integrals." With today's ease of computation, you may find it interesting and worthwhile to look at the literature on the subject and to replicate and extend some of the computations.

2.28 Poincaré maps

To investigate the model more generally, work with Poincaré maps. Start with small μ: 0.01 or less. Planetary-type orbits around the primary will produce points lying on well-defined invariant curves. At the center of a system of invariant curves there is an invariant point, corresponding to a (nearly) circular orbit around the primary. Now increase μ. Notice how the character of the maps changes and how the invariant curves and their central invariant point disappear. A conclusion is that orbits of planetary type in a binary system can only exist when the mass of one member is much smaller than the mass of the other.

2.29 Non-stellar applications
Don't neglect other possible applications. For instance, with $\mu = 1/81$ you can investigate orbits in the Earth-Moon system.

The Photo-Gravitational Problem of Three Bodies

2.30 Compare with previous results
Some of the preceding exercises should be replicated and enlarged upon with the introduction of radiation pressure.

If the primary is a white dwarf, then radiation pressure need only be considered from the secondary star. Introduce this gradually, until the figure-eight curve of zero velocity opens out beyond the primary. As before, investigate orbits passing through L_1. If the orbit streams form a hot spot, then the mechanism for forming an accretion disk may be applied. For various values of μ find the radiation pressure parameter when the hot spot is no longer formed.

2.31 Chaotic motion
Investigate whether radiation pressure affects the chaotic and non-chaotic motions.

2.32 Radiation from both stars
Generalize the exercises for cases where both stars are exerting radiation pressure.

Refer to Simmons et al.[11] Find conditions where L_1 is stable. Confirm this by observing orbits close to L_1.

The Formation of an Accretion Disk

2.33 Use of the program on interacting galaxies
The program for two interacting galaxies, following the motion of up to 500 particles in the field of two gravitation bodies, can be used to consider possible outer boundaries for the disk. The first galaxy becomes star A and the second, star B. Choose initial conditions so that B is in a circular orbit with zero inclination. For different mass ratios, investigate the largest disk size that may be possible dynamically.

2.34 Light curves if the stars are observed to eclipse
Consider some light curves for eclipsing binaries. How would they be modified by the presence of a hot spot?

2.35 Application involving L_1
Included in the program are procedures for locating L_1 and the boundaries of the figure-eight. Either write your own programs or use a hand

calculator with a solver. Check results against those of the program. These might be used separately in various projects. Examples:

a. Given the stellar masses and the orbital period, find the separation (assuming circular motion) and locate L_1. Approximate the lobe around the secondary by a sphere and find the mean density of the star, assuming it to fill the lobe. Compare this with densities of stars of different types.

b. As mass is transferred from one star to the other, with the total angular momentum conserved, the separation and period are changed such that

$$\frac{\Delta A}{A} = \frac{2\Delta m(m_1 - m_2)}{m_1 m_2} = \frac{2}{3}\frac{\Delta P}{P},$$

where Δm is the mass lost by m_2 and gained by m_2 and ΔA and ΔP are the changes in separation and period, respectively. See if you can justify this result.

c. It has been shown (see the article by A. R. King,[4]) that if $m_2 > m_1$, then a loss of mass in this way moves L_1 toward the secondary; so in this case effects of mass loss will be accelerated. However, we actually find $m_2 < m_1$ when L_1 moves away. See if you can establish this result. Try running the preceding program sequentially, with allowance for loss of mass from the secondary.

The problem of how the secondary is kept in contact with its Roche lobe has received much attention. You should explore the literature on the subject.

References

1. Aitken, R. G. *The Binary Stars*. New York: Dover Publications, 1964.

2. Batten, A. H. *Binary and Multiple Systems of Stars*. New York: Pergamon Press, 1973.

3. Binnendijk, L. *Properties of Double Stars*. Pennsylvania: University of Pennsylvania Press, 1960.

4. Bode, M. F., Evans, A. ed. *Classical Novae*. New York: Wiley, 1989.

5. Couteau, P. *Observing Visual Double Stars*. Cambridge: MIT Press, 1981.

6. Danby, J. M. A. *Fundamentals of Celestial Mechanics*, 2nd ed. Richmond: Willmann-Bell, 1988.

7. Hut, P. Tidal Evolution in close binary systems. Astronomy and Astrophysics **99**:126–140, 1981.

8. Kopal, Z. *Dynamics of Close Binary Systems*. Dordrecht, Holland: J. Reidel, 1978.

9. Poincaré, H. *Les Méthodes Nouvelles de la Méchanique Célests*. Paris: Gauthier-Villars, 1892.

10. Pringle, J. E., Wade, R. A. eds. *Interacting Binary Stars*. Cambridge, UK: Cambridge University Press, 1985.

11. Simmons, J. F. L., McDonald, A. J. C., Brown, J. C. The restricted three body problem with radiation pressure. Celestial Mechanics **35**: 145–187, 1985.

12. Smart, W. M. *Textbook on Spherical Astronomy*, 6th ed, revised by Green, R. M., Cambridge, UK: Cambridge University Press, 1977.

3

On the Motion of N-Bodies

J. M. Anthony Danby

3.1 The Model of Wright and the Toomres

3.1.1 The Model

This program follows the model proposed by A. E. Wright[10] and A. and J. Toomre[9] for investigating the interaction between two galaxies. Each galaxy is taken to be, gravitationally, a point mass, and the "stars" are massless particles. Each star, therefore, is following an orbit in the restricted problem of three bodies, since there is no gravitational action between the stars. The seminal papers have inspired a considerable amount of research, and numerical experimentation, as well as impressive animated displays. For up to a few hundred stars, the animation can be viewed in real time on the computer, so that experiments with varying the parameters of the model can be performed quickly and easily. If the speed is too slow, then use fewer stars. In running the program, you will have the option of storing the animation to a file, which can be played back later at an increased speed.

In the program, as it stands, the stars are initially in rings, moving in coplanar circular orbits around the galaxies. The second galaxy moves relative to the first in a Keplerian orbit, the elements of which can be specified. The animated motion can be viewed from a direction perpendicular to the initial plane of the stellar orbits (the x-y plane); it can also be viewed simultaneously as projected in the x-y and x-z planes. A star is moved immediately after its new position is computed, leading to a more active appearance. During a close approach of a galaxy and a star, more intensive computation is needed; so, although the motion becomes faster in real time, it will slow down in the display. If the motion is stored onto a file, then it is played back, frame by frame, in real time. (In a playback the motion can be viewed from any angle.) Do not begrudge the extra time taken by close encounters; this is where the significant physics is taking place.

3.1.2   Units

For a program of this kind, it is essential that canonical units be used so that all the numbers involved are of moderate order. In the program the mass of the galaxy A is **one** unit of mass, and the constant of gravitation is equal to **one**. The output screen is scaled so that distances from A should be 5 units at most. Change to actual units will depend on the application considered. For the galactic model, suppose the mass of A to be 10^{11} solar masses, and one unit of length to represent 10 kpc (kiloparsec). Consider a star moving in a circular orbit one unit from A. By Kepler's third law, its period, in years, would be

$$P = \frac{2\pi}{365.25} \sqrt{\frac{(206265 \times 10^4)^3}{10^{11}k^2}},$$

where $k = 0.017202$ is the Gaussian gravitational constant. There are 206,265 astronomical units in a parsec. This simplifies to

$$P \approx 3 \times 10^8 \text{ years}.$$

So 2π units of the *program's time* correspond to 3×10^8 years. For an application with the solar system, "galaxy" A might become the Sun, with the unit of length 100 astronomical units; then the period of a circular orbit with unit radius would be

$$P = \frac{2\pi}{365.25} \sqrt{\frac{(10^2)^3}{k^2}} = 10^3 \text{ years},$$

equivalent to 2π units of the program's time.

3.1.3   Method of Computation

The computational methods have been selected, and some compromises have been made, in order to maximize the speed of execution. Several parameters are used, and can be varied by changes in the code. These are set numerically at the start of the program.

The galaxy A is located at the center of the screen; the relative Keplerian orbit of B is computed using **f** and **g series** (see Danby,[4] section 6.7) and also section 9 in the Appendix to this volume. This provides fast computation of the location of B as it is needed in the computation for each separate star. **But it will become invalid for close approaches of A and B, and will lead to errors if they collide.** When the force on a star is to be calculated, the number of terms used in the series for galaxy B is "NMaxS;" (the default value is 3). When computing the position of galaxy B at the end of a time step the number of terms used is "NMaxG;" (the default value is 6). The time interval between plotted position is "dt;" (the default value is 0.5). The series may have to be reinitialized during this interval; the maximum allowable error in the series computation for the position of galaxy B is "TruncError;" (the default value is 0.00001). The maximum time interval for the use of a series is determined from this maximum error; if the maximum time is less than *dt*, the series must be reinitialized.

The orbit of each star is computed by numerical integration using a Runge-Kutta-Fehlberg method with variable stepsize of order three (for details, see

E. Fehlberg[5]). The variable stepsize is controlled by the condition that the local truncation error of any step must not exceed a given tolerance, taken in the program to be 0.00001. The integration step is taken in the Procedure StarMove. As soon as the position of a star has been found, the star is moved. This leads to more continuous appearing animation.

If a star approaches a galaxy within the distance "TooClose," then the attraction of the second galaxy is ignored, and the star is "thrown" around the galaxy in a Keplerian orbit; in this operation universal variables are used. This procedure will break down if A and B are both within the distance 0.1 from the star. The default value of TooClose is 0.05.

The inverse square law of attraction is "softened" if a star approaches a galaxy within a distance $r < \epsilon$ then $1/r^2$ is replaced by $1/\epsilon^2$. ϵ is represented by "Eps" in the program. If the Keplerian approximation just mentioned is operative, then there is no cause to activate this softening.

If you elect to store the output in a file, the program opens a file named by you. This can be viewed using the program PLAYBACK, described below.

3.1.4   Suggested Changes to the Program

1. The orbits of the stars need not lie in a plane.

   Suppose that the initial conditions are to be contrived so that the stars are, approximately, spherically symmetrical around galaxy A. With origin at A take the spherical coordinates θ and ϕ. θ should be a random number between 0 and 2π, and $\sin \phi$ should be a random number between 0 and 1. $\cos \phi = (1 - \sin^2 \phi)^{1/2}$ for positive z; to achieve randomness with positive or negative z, its sign should be changed randomly.

   For distance r from the center of A, the coordinates of the star are

   $$(r \cos \theta \sin \phi, \text{ r} \sin \theta \sin \phi, \text{ r} \cos \phi).$$

   If the circular speed is v_c, then the velocity components would be

   $$(-v_c \sin \theta, v_c \cos \theta, 0).$$

   The distribution could be made spheroidal instead of spherical by scaling the z-position components.

   This model might be used to investigate the effects of a passage on an elliptical galaxy.

   (Note: You can see an example of the random choice of initial conditions in the program for the motion of N attracting bodies.)

2. Allowance might be made for unseen mass.

   This could be done most easily by enclosing the entire mass inside a uniform spherical distribution of matter. Then you will need to include a term $-A\vec{r}$ on the right-hand sides of the differential equations. For the stars, this could be achieved by adding the line

   z[n+3] := z[n+3] + A*rStar[n];

   in the Procedure Fun, which is part of the larger Procedure Advance.

The **f** and **g** series now satisfy the differential equations

$$\left. \begin{array}{l} \dfrac{d^2\mathbf{f}}{dt^2} = -\mu\dfrac{\mathbf{f}}{r^3} - A\mathbf{f}, \\[2ex] \dfrac{d^2\mathbf{g}}{dt^2} = -\mu\dfrac{\mathbf{g}}{r^3} - A\mathbf{g}. \end{array} \right]$$

So the Procedure FandG should include, immediately after the end of the loop (**k := 0 TO n−2**), the instructions

f[n] := f[n] − A∗f[n−2];   and   **g[n] := g[n] − A∗g[n−2].**

3. Additional or different forces might be included.

   For example, if the "galaxies" are interpreted as members of a binary system, at least one of the stars might be exerting radiation pressure that would affect the "stars," or particles in the disk, but would not affect the other star. This effect could be included by a simple modification of the Procedure Fun.

4. The galactic orbits might decay.

   In their paper the Toomres include the following paragraph:

   > When two roughly equal disk galaxies in a previously long-period orbit at last experience a close approach and raise really violent tides in each other, can it be that they also surrender a significant fraction of their orbital energy (and angular momentum) to the tail-making particles, many of which of course escape to infinity? And hence would not their remnants drop into orbits of progressively shorter periods, until at last they lose altogether their separate identities and simply blend or tumble into a single three-dimensional pile of stars?

   These questions cannot be answered by the use of a model that denies the component stars any mass, and hence any energy to steal from the two galaxies. However, suppose that they had **some** mass, but not enough to influence one another. Suppose that you banish from the model any star that strays further than some definite distance from galaxy A, and at the instant of banishment, you calculate its orbital energy and angular momentum, and subtract them from those of the galaxies. You should see a decay in their orbits. This is not a clear-cut activity, since the mechanics are faulty. But you may discover some suggestive results.

5. The mass of one of the galaxies could be artificially reduced or increased during a run. This could be a possible mechanism for artificially introducing orbital decay.

6. Another interpretation of variable mass is as follows. Think of the first galaxy as the Sun and the second as Jupiter. The rings of stars would be initially in orbit around Jupiter, forming a system of satellites. The project

would be to investigate the survival of satellite systems; they might be dissipated through solar action and also if Jupiter's mass were to diminish—as may have happened in the early history of the solar system.

3.1.5 Running the Program

The relevant items in the menu for running the program are

Data   Orbits

Selection of **Data** produces an input screen in which the following information is to be entered:

"Store results in a file? (Y or N)"

"File name:"

"Enter the number of rings of stars (10 or fewer)"

Hit **ENTER** to proceed to the next input screen. This will ask you for values of the radius of each ring and the number of stars in each ring. Remember that the total number of stars cannot exceed 500. The graphics screen is scaled for a range of ±5 units of distance.

The third input screen prompts you for data on the second galaxy—its mass and orbital elements. The mass of galaxy A is programmed to be one unit of mass. The line of nodes will be horizontal on the screen. You can also choose to have up to five rings of stars around the second galaxy. If you do so, there will appear a final input screen for the radii and numbers of stars in each ring; you will also be asked for two angles, Theta and Phi, specifying the orientation of the plane of these rings. The plane of the rings intersects the x-y plane in a line making the angle Theta with the x-axis, and is inclined to the x-y plane by the angle Phi.

When the data are complete, you have the choice, through **Orbits**, of seeing the animated motion in the initial orbital plane of the stars around galaxy A, the x-y plane, or, in a split screen, the x-y and x-z planes. The animation can be controlled by hot keys.

3.2 Application to the Asteroids

3.2.1 The Model

In principle, this program is the same as the preceding one; it was conceived as an illustration of how that program could be adapted to a special ambience. In detail, however, there are considerable changes. The question addressed here has to do with orbital resonance, and specifically resonance of the orbit of an asteroid with that of Jupiter. It is well known that there are gaps in the system of asteroids corresponding to some resonances (2:1, for instance) but that other resonances are occupied by groups: 1:1 for the Trojans or 3:2 for the Hilda group (which includes 3415 Danby!). Running this program may not provide significant answers to the problems posed by these resonances; but it will enable you actually to see the orbits of the asteroids with a selected resonance and to follow the perturbation in their orbits—perturbations that may throw the asteroids out of the initial resonance and perhaps into the path of a planet such as Mars.

The name "Jupiter" is used symbolically as the "dominant planet" in the system. You will prescribe its mass and orbital elements.

The maximum allowable number of asteroids is 100. The animation is of less significance than the perturbation analysis: the latter provides the bulk of the added code. Also, notation has been changed so that the variables reflect the Sun-Jupiter system rather than galaxies.

The perturbations that are calculated and plotted are the *osculating elements*. If, at any instant, the perturbing force due to Jupiter were to disappear, then the asteroid would move in a Keplerian orbit with the Sun as focus. The elements of that orbit are the osculating elements at that instant. Thus the osculating elements are slowly varying quantities (usually) that provide some idea about the geometrical nature of the perturbations. In the Procedure SeeElements an asteroid is selected by clicking on it, and the orbit of that asteroid is integrated numerically in Cartesian coordinates. At each step of integration the osculating elements are calculated and stored for later plotting.

3.2.2   Units

The unit of mass is the solar mass and the unit of distance is the astronomical unit. The constant of gravitation is taken to be equal to one, so that, by Kepler's third law, one year would be equal to 2π units of the program's time. In the displayed output, the unit of time is the year.

3.2.3   Suggested Changes to the Program

1. As before, the system could be embedded in unseen mass. A more interesting change is to decrease, with time, the mass of the Sun; this could correspond to conditions in the early stages of a planetary system. Starting with a system of asteroids inside the orbit of Jupiter, could these be captured into a resonance? In particular, could the presence of the Trojan asteroids be explained in this way?

2. The "asteroids" might be interpreted as particles that are subject to Poynting-Robertson perturbations. Let ϵ be a parameter that is essentialy equal to the ratio of the radiation force divided by the gravitation force. Then if a particle has position and velocity $\vec{r}$ and $\vec{v}$ relative to the Sun, the radial force, per unit mass, on the particle becomes $-GM(1 - \epsilon)\vec{r}/r^3$ and the Poynting-Robertson drag force is $-(GM\epsilon/c)\vec{v}/r^3$. Here G is the constant of gravitation, M is the mass of the Sun, and c is the speed of light. To detect effects in a numerical experiment, the value of c may be varied. Orbits of particles in general decay toward the Sun; but it has been claimed that some, especially particles starting outside the orbit of Jupiter, may become trapped into a resonance.

3. The parts of the program for computing and plotting osculating elements might be taken to form a utility to be applied to more general problems involving perturbed Keplerian motion. These might include perturbations of a lunar-type orbit, of a satellite of an oblate planet, or a satellite due to atmospheric drag. An interesting dynamical model involved a perturbation having constant magnitude and direction.

4. It can be instructive to view the animation of the asteroids in a rotating reference frame. There are two principal possibilities. The axes might rotate with the mean motion of the asteroids; you will see the libration in their orbits. Or the axes might rotate with Jupiter.

3.2.4 Running the Program

The items in the menu that control the running of the program are

Data   **See Orbits**   **Asteroid**   **Elements**

Data produces two input screens. In the first, you are prompted to enter the mass of Jupiter (in solar masses) and its orbital data. It is assumed to have zero inclination. In the second screen you enter numbers p and q to define a resonance $p{:}q$. This means that p revolutions of the asteroid take the same time as q revolutions of Jupiter. Next, you can have a scatter in the initial eccentricities of the orbits of the asteroids. They will be chosen at random between zero and De, where you are prompted for a value of De. In the same manner, you are prompted for a value of Di, for the inclination. Perihelia, nodal angles, and initial true anomalies are picked at random by the program to lie between 0 and 2π. It might be instructive if you were to see a reference circle during the animation; it could represent the orbit of Mars. You can choose to have this, and can select the radius. Finally, you are asked if you want the animation to be stored in a file; if so, name the file.

See Orbits shows the animated motion of Jupiter and the asteroids. It can be interrupted by—**Asteroid**. You are prompted to select any asteroid by clicking on it. Further results will apply to the orbital history of that asteroid. Having selected the asteroid, you will be prompted for a time interval over which its orbit is to be integrated. During the integration the osculating semimajor axis will be plotted. The default range for this plot is the value corresponding to the input resonance plus or minus Da, with $Da = 0.3$. You can, if you like, change Da. The value of the semimajor axis corresponding to the resonance is marked by a horizontal line across the screen.

Once the plot is complete, you can use the items in **Elements** to see the plot of any element. (If the inclination is zero, the inclination and nodal angles will not be accessible.) You can also extend the time for further integration.

As set up, the plotting of osculating elements may produce strange results if the perturbations are very large.

3.3 *The Motion of N Attracting Bodies*

The program described in this section is intended to be a utility for the numerical integration of the motion of n attracting bodies. The number of bodies can be selected by the operator, as can the motive for running the program. In this section and the next two possible applications are described. To make others, you need to decide how you want your initial conditions to be defined and what you want to see from running the program. The program might be generalized; for instance, the system could be "immersed" in unseen mass. This sort of development, however, I leave to you.

As it is set up here, the motive is to observe the possible validity of the virial theorem. Therefore, we shall start with a brief exposition.

3.3.1 The Virial Theorem

From Eq. (A.2) we have the equations of motion of n bodies to be

$$m_i \frac{d^2\vec{r}_i}{dt^2} = G \sum_{\substack{j=1 \\ j \neq i}}^{n} m_i m_j \frac{\vec{r}_j - \vec{r}_i}{\|\vec{r}_j - \vec{r}_i\|^3}. \tag{3.1}$$

The potential energy of the system is

$$V = -G \sum_{i<j}^{n} \sum_{j=1}^{n} \frac{m_i m_j}{r_{ij}}. \tag{3.2}$$

Then

$$\frac{\partial V}{\partial x_i} = -G m_i \frac{\partial}{\partial x_i} \left\{ \sum_{j=1}^{n} \frac{m_j}{r_{ij}} \right\}$$

$$= G m_i \sum_{j=1}^{n} m_j \frac{x_i - x_j}{r_{ij}^3}.$$

Therefore, Eq. 3.1 can be written

$$m_i \frac{d^2\vec{r}_i}{dt^2} = -\nabla_i V, \tag{3.3}$$

where

$$\nabla_i \equiv \hat{\imath} \frac{\partial}{\partial x_i} + \hat{\jmath} \frac{\partial}{\partial y_i} + \hat{k} \frac{\partial}{\partial z_i}.$$

Multiply Eq. 2.2 scalarly by $\dfrac{d\vec{r}}{dt}$ and add all the n equations; we have

$$\sum_{i=1}^{n} m_i \frac{d\vec{r}_i}{dt} \cdot \frac{d^2\vec{r}_i}{dt^2} = -\sum_{i=1}^{n} \frac{d\vec{r}_i}{dt} \cdot \nabla_i V$$

$$= -\frac{dV}{dt}.$$

Integrating this, we find

$$\frac{1}{2} \sum_{i=1}^{n} m_i \left(\frac{d\vec{r}_i}{dt} \right)^2 = -V + \text{constant}, \tag{3.4}$$

which is the energy integral. Let T be the kinetic energy of the system; then

$$T = \frac{1}{2} \sum_{i=1}^{n} m_i \left(\frac{d\vec{r}_i}{dt} \right)^2,$$

and the energy integral can be written

$$T + V = h. \tag{3.5}$$

Next we define the quantity I, akin to a "moment of inertia," as

$$I = \frac{1}{2} \sum_{i=1}^{n} m_i \vec{r}_i^2. \tag{3.6}$$

Differentiating it twice with respect to the time, we find

$$\frac{d^2 I}{dt^2} = \sum_{i=1}^{n} m_i \left(\frac{d\vec{r}_i}{dt} \right)^2 + \sum_{i=1}^{n} m_i \vec{r}_i \cdot \frac{d^2 \vec{r}_i}{dt^2}$$

$$= 2T - \sum_{i=1}^{n} \vec{r}_i \cdot \nabla_i V.$$

Now V is a homogeneous function of all the coordinates of order -1. A function $f(x_1, x_2, \ldots, x_n)$ is called homogeneous, of order k, if

$$f(\lambda x_1, \lambda x_2, \ldots, \lambda x_n) \equiv \lambda^k f(x_1, x_2, \ldots, x_n). \tag{3.7}$$

If we differentiate this identity with respect to λ, and then set $\lambda = 1$, we have

$$\sum_{i=1}^{n} x_i \frac{\partial f}{\partial x_i} = kf. \tag{3.8}$$

This is Euler's theorem on homogeneous functions. Then

$$\sum_{i=1}^{n} \vec{r}_i \cdot \nabla_i V = \sum_{i=1}^{n} \left\{ x_i \frac{\partial V}{\partial x_i} + y_i \frac{\partial V}{\partial y_i} + z_i \frac{\partial V}{\partial z_i} \right\}$$

$$= -V.$$

Therefore,

$$\left. \begin{array}{l} \dfrac{d^2 I}{dt^2} = 2T + V, \\[2mm] \qquad = 2h - V, \\[2mm] \qquad = T + h, \end{array} \right] \tag{3.9}$$

where we have used the energy integral, Eq. 3.5. This constitutes the virial theorem.

Some astronomical assemblages of bodies, such as stellar associations, dissipate rapidly; but many systems do not change substantially over long periods of time; they are said to be in a *statistically steady state* if I is constant, so that

$$2T + V = 2h - V = T + h = 0. \tag{3.10}$$

But suppose that the system may be evolving with time. Take the *time average* of all terms in the virial theorem from $t = 0$ to $t = \tau$. This is written as

$$\frac{1}{\tau}\int_0^\tau (2T + V)dt = \frac{1}{\tau}\int_0^\tau \frac{d^2I}{dt^2}dt$$

$$= \frac{1}{\tau}\left[\frac{dI}{dt}\right]_0^\tau$$

$$= \frac{1}{\tau}\left[\sum_{i=1}^n m_i\vec{r_i} \cdot \frac{d\vec{r_i}}{dt}\right]_0^\tau .$$

Then if I and T remain bounded, it is clear that the right-hand side approaches zero with increasing τ. Further, the limits

$$\langle 2T \rangle = \lim_{\tau \to \infty} \frac{1}{\tau}\int_0^\tau T(t)dt, \quad \text{and} \quad \langle U \rangle = \lim_{\tau \to \infty} \frac{1}{\tau}\int_0^\tau U(t)dt$$

exist, and

$$\langle 2T \rangle + \langle V \rangle = 2h - \langle V \rangle = \langle T \rangle + h = 0 . \qquad (3.11)$$

It is of interest that H. Pollard[8] has established the stronger result: The statement $\langle T \rangle + h = 0$ is true if and only if

$$\lim_{\tau \to \infty} \tau^{-2}I(\tau) = 0 .$$

So boundedness is not a necessary condition.

3.3.2   The Program

The program numerically integrates the equations of motion (Eq. 3.1), of a selected number of bodies in an entirely unsubtle way, using a fifth-order Runge-Kutta-Fehlberg[5] method with variable stepsize. This means that the animation will slow down during close encounters. But the animation can be stored for later playback in real time. Units are unspecified; however, the gravitational constant is taken to be equal to one. The heart of the program resides in the Procedure Fun, which is called by the numerical integrator. In this section and in the one that follows, two specific applications are described.

Several parameters are controlled by changing quantities in the program's code rather than entering them on an input screen. These parameters are as follows:

1. The maximum number of bodies (currently 20). This is **MaxNBodies**. The parameter **MaxNeq** should be set to be six times **MaxNBodies**.

2. The time interval between plotted positions, **Dt**, of bodies (currently 0.1).

3. The distance, **Eps**, within which the force of attracton is taken to be constant. This controls the "soft" gravitational attraction. If this were zero, then in a close encounter the program will slow down drastically or crash with an "illegal floating-point operation" message.

4. The masses of the bodies. Currently they are all equal. These are stored in an array **Mass[n]** and are initialized in the Procedure InitialConditions.

There are three functions: Potential, which finds V; KineticEnergy, which finds T; and InertiaFunction, which finds I. The motion is displayed in animated form, and the current values of $2T$, V, $T + V$, and I are displayed. Since the program includes a provision of a "soft" inverse square force (where for distances of approach smaller than a specified amount, the force is constant), the quantity $T + V$ can have occasional disconcerting fluctuations during close approaches. Apart from these aberrations, the quantity $T + V$ should remain constant; a change in this number indicates integration truncation error.

Seeing the virial theorem in action is harder than you might suppose, because even with negative total energy, there is a high probabiliy that a body will escape. A facility that is intended to take this into account is under the **Remove** option in the menu. If a body has positive total energy and is beyond a given distanced from the center of mass of the system, then it is removed from the system; after this, parameters for the remaining bodies are followed. The starting scale for graphics has limits ± 2; the escape distance is 3 units in this scale.

You will be prompted to enter the number of bodies and the total energy, h, of the system. Initially, the masses are distributed at random within a sphere of radius one unit of length. As has been mentioned, the masses have all been set equal, but the program is easily changed to allow unequal masses. The potential energy is computed. Then a mean speed is found from

$$\bar{v} = \sqrt{\frac{2(-V + h)}{\sum m_i}}$$

(which is where you may get into trouble if h is too small, so that $-V + h < 0$). The bodies are given this speed but in random directions.

3.3.3   Suggested Alterations to the Program

1. The nature of the input can be changed. In particular, the randomness can be removed so that definite configurations can be followed, and altered in a systematic way.

2. The system might be embedded in a surrounding field of force. The simplest way to do this would be to surround the system by a uniform spherical distribution of matter. The generalization of the equations of motion is simple. You should also appropriately modify the statement of the virial theorem.

3. Modify the inverse square law of force. If you want to follow the existing displayed functions, these will have to be modified also.

3.4   *Make Your Own Solar System*

This is an application of the utility introduced in the preceding section. The model consists of the Sun, "Jupiter," and up to nine other planets. The reference plane is that of the initial orbit of Jupiter. The masses and orbital elements are specified by the operator. **Units.** The unit of mass is the mass of the Sun, and the unit of length is the astronomical unit. The constant of gravitation is equal to one, so that one year is equivalent to 2π units of the program's time.

3.5 Playback

PLAYBACK is a utility that runs the animation from any of the files stored in the four programs just described. Each of these files will be of the form YOUR-NAME.POS. When you select a file, you will also be prompted for the time delay between the showing of successive frames.

Once a file has been selected, the animation can be viewed either in a single screen or in split screens. In the first case the motion is projected into the *x-y* plane (as is seen in the original program). In the second case, the screen shows that same projection; but the viewing direction in the right-hand screen can be varied by the use of the **arrow** keys; a set of axes is also shown in order to indicate the current direction of viewing.

3.6 Exercises

Program on Interacting Galaxies

3.1 **Changes in parameters**

Read the original papers by Wright[10] and the Toomres,[9] and look for more recent papers dealing with similar material (see, for instance, Barnes[2]). In these papers circumstances were looked for that would produce "bridges and tails" so that comparison might be made with observed galaxies that appear to be interacting.

The principal parameters of the problem are the mass of the second galaxy, its perigalacticon distance, and the eccentricity and inclination of its orbit. You should investigate the effects of changing these parameters systematically, using a limited number of stars.

Find some illustrations of observed interacting galaxies and look for varieties of parameters that might account for these. See Arp[1]. For more information on galactic structure, see Mihalas and Binney.[7]

3.2 **Solar system and comets**

Interpret galaxy A as the Sun, the stars as members of the Oort cloud of comets and the second galaxy as the problematical star Nemesis that, perhaps, with periodic close approaches to the Oort cloud causes unusually many cometary orbits to be deflected toward the inner part of the solar system. The problem here is to look for parameters (if they exist) that make it possible for the cloud to be perturbed sufficiently for comets to approach within one astronomical unit of the Sun, but without destroying the cloud over many passages.

3.3 **Sun, Jupiter and asteroids**

Interpret galaxy A as the Sun, galaxy B as Jupiter, and the stars as asteroids. Set Jupiter in a circular orbit around the Sun, and start with rings of asteroids including one with a commensurability, such as 2:1 (so that an asteroid revolves around the Sun twice while Jupiter revolves once), the others being close by. One of the Kirkwood gaps is at the 2:1 commensurability. See if you can gain some insight into this.

3.4   Sun, Jupiter and satellites

Interpret galaxy A as Jupiter, galaxy B as the Sun, and the stars as satellites of Jupiter. Under what circumstances can Jupiter hold on to its satellites? It has been suggested that the Trojan asteroids were once satellites of Jupiter that escaped; why is this not feasible?

3.5   Rings of a planet

Interpret galaxy A as Saturn, galaxy B as a satellite of Saturn, and the stars as ring particles. The mission here would be not so much to try to represent the actual rings, but to find situations when rings might be possible or impossible, or where gaps might appear.

3.6   Modeling an accretion disk

Interpret the "galaxies" as members of a binary system and the "stars" as particles in an accretion disk around one of the stars. This is, strictly, part of one of the other projects; but the present program is needed. In a close binary the relative orbits can be taken to be circular. If there is an accretion disk around one of the stars, then it will lie in the orbital plane. The orbits within the disk should be nearly circular. For different mass ratios of the two stars, what maximum radius might be reasonably expected for the disk? (Usually this is discussed within the framework of the circular restricted problem of three bodies; the unit of mass is the sum of the two masses, and the unit of distance is the distance between the two bodies. This should be noted if comparison is made with the program dealing with accretion disks.)

3.7   Planets in a binary system

Investigate the question: "Can a binary star system support a planetary system?" The two galaxies would become the binary stars; the stars would become planets. How far outside the binary would a ring of planets need to be placed in order for them to remain in the system?

Application to the Asteroids

3.8   Different types of motion

One object in working with a program such as this is to try to account for the properties of a system of asteroids as it exists today. You have the freedom to make up your own system, and to see how it might evolve. There has been a considerable amount of experimental work done on imaginary asteroids. Some references are given below. In particular, see *An Introduction to Asteroids* by C. J. Cunningham.[3] Become familiar with some of the experiments and results that have been claimed, and see if you can replicate any of them. Other helpful references include Hamilton and Burns,[6] and Yoshikawa.[11]

You can observe orbital motion of a ring of asteroids or the perturbation of the elements of a single asteroid. Look for cases where asteroids may be expelled from a commensurability, leaving a gap. Also look for evidence of chaotic motion. When observing the elements, concentrate on the semi-major axis and eccentricity as being most significant. Quasi-periodic motion should appear fairly clearly in the changes of these elements, as should chaotic motion.

3.9 Chaotic motion in different resonances

Zones of chaotic motion have been found close to the 2:1 and 3:2 resonances. It has been claimed that close to the 2:1 resonance the chaotic orbits will cross the orbit of Jupiter but that this is not the case for the 3:2 resonance. It has also been claimed that collisions with the planets may have removed those asteroids with chaotic orbits within the lifetime of the solar system. These are typical of some of the arguments that you might investigate.

For seeing the perturbations of orbits close to but not at an initial resonance, use nearby fractions; for instance, 13:6 might be considered "close" to 2:1.

3.10 Effects of Jupiter's mass

In making up your own system of asteroids you are free to give Jupiter any mass that you please.

The main parameters of this model are the mass of Jupiter and the initial resonance; the effects of altering other parameters can be investigated secondary to the first two. For moderate values of Jupiter's mass you will see quasi-periodic motion, built up from a combination of different frequencies. This can be observed by watching the perturbation of the elements. Compare the amplitudes for different resonances. Then gradually increase the mass of Jupiter and see what effects this may have. The restricted problem is an example of a chaotic dynamical system. The onset of chaos involves the destruction of apparently quasi-periodic behavior. If the mass of Jupiter is high enough, then the existence of a system of asteroids will become impossible. Can you confirm this?

3.11 Asteroids outside the orbit of Jupiter

The observed asteroids have orbits lying between Mars and Jupiter. The program permits you to locate them outside Jupiter's orbit. Can you find significant resonances in this region? (To discuss this further, you need to include Saturn in the model. You can do this in the program PLANET: Make Your Own Solar System.)

3.12 Trojan asteroids

The program is not designed to give reasonable results if the asteroid passes close to Jupiter; momentarily, the osculating elements (which are calculated relative to motion around the Sun) may go haywire. Also the scaling of their plots depends on their extreme values, and this can ruin, in particular, the plot of the osculating semimajor axis. With this warning, I suggest that you experiment with the 1:1 resonance. With a uniform circular distribution to start with, you may end up with some satellites of Jupiter and two set of Trojan asteroids. Try it!

The Motion of *N* Attracting Bodies

3.13   **Stability of three bodies**
The principal application of this utility is to gain insight into *n*-body motion. Of prime importance are collisions and escapes. One aproach is to start with $n = 3$. Vary the energy. Find out how unlikely it is that the system will remain bounded. If it does remain bounded, find out the usual final configuration.

3.14   **Stability of four bodies**
Next, with $n = 4$, repeat the experiment.

3.15   **Effects of energy on stability**
For higher values of *n*, experiment with the energy. Observe that systems having positive total energy may dissipate very rapidly. See if you can set up a demonstration showing the validity of the virial theorem. (It is not easy!)

3.16   **Generalization of the virial theorem**
Generalize the virial theorem to allow the addition of a gravitational field due to a spherical distribution of matter.

Make Your Own Solar System

3.17   **Increasing Jupiter's mass**
One question that might be asked is, "If Jupiter's mass were increased, could our solar system remain the same?" Would, for instance, "proto-" giant planets with ten times their present masses have allowed the other planets to assume and retain their present orbits? The time scales involved might defeat the thorough investigation of some of these questions; but you can be fairly certain that if the motion is at all tempestuous, then eventually some bodies will be expelled. (But the "soft" gravitation that is used will reduce the severity of close encouters and the number of expulsions.)

3.18   **Jupiter, Saturn and asteroids**
Some of the masses can be zero. You might look into such questions as, "Could an asteroid system exist between the orbits of Jupiter and Saturn?

3.19   **Asteroid resonances**
Consider a system consisting of Jupiter, Mars, the Earth, and some asteroids. Start the asteroids in resonant orbits. It has been claimed that perturbations can "pump up" the eccentricities so that the asteroids may encounter the other planets. Can you replicate this?

3.20   **Planetary resonances**
Jupiter and Saturn have orbits close to a 5:2 resonance. What would happen if this were to be precise? How about a 2:1 resonance?

3.21 **Application to stellar systems**

Although the word "planets" is used, the masses are at your discretion. You could experiment with two stars and a possible planetary system, or a group of stars. Stellar associations usually dissipate; survivors tend to group in pairs. Can you observe this?

References

1. Arp, H.C. *Atlas of Peculiar Galaxies*. Pasadena: California Institute of Technology, 1966.

2. Barnes, J.E., Hernquist, L. Interacting galaxies. Annual Review of Astronomy and Astrophysics **30**:705–742, 1992.

3. Cunningham, C.J. *Introduction to Asteroids*. Richmond: Willmann-Bell, 1988.

4. Danby, J.M.A. *Fundamentals of Celestial Mechanics*, 2nd ed. Richmond: Willmann-Bell, 1988.

5. Fehlberg, E. Low order classical Runge-Kutta formulas with stepsize control. NASA TR 315, 1969.

6. Hamilton, D.P., Burns, J.A. Orbital stability zones about asteroids. Icarus **92**:118–131, 1991.

7. Mihalas, D., Binney, J. *Galactic Astronomy. Structure and Kinematics*. San Francisco: W.H. Freeman and Company, 1981.

8. Pollard, H. *Mathematical Introduction to Celestial Mechanics*. New Jersey: Prentice-Hall, 1966.

9. Toomre, A., Toomre, J. Galactic bridges and tails. Astrophysical Journal **178**:623–666, 1972.

10. Wright, A.E. Computational models of gravitationally interacting galaxies. Monthly Notices of the Royal Astronomical Society **157**:309–333, 1971.

11. Yoshikawa, M. *Motion of asteroids at Kirkwood gaps*. Icarus **92**:94–111, 1991.

4

Galactic Kinematics

J. M. Anthony Danby

4.1 The Rotation Curve of a Galaxy

A spiral galaxy is concentrated toward a plane and much of its material, gaseous clouds in particular, moves in circular orbits about the galactic center. If $V(R)$ is the circular velocity at a distance R from the galactic center, then the function $V(R)$ is called the *rotation curve* of the galaxy. In principle, the rotation curve of a galaxy can be observed spectroscopically. In the program to be described, you can make up your own galaxy, observe its rotation in animation, and see the rotation curve. For a given rotation curve (either real, or constructed by the program), you might also try to reconstruct a model for the galaxy.

In the program the galaxy is constructed from a central point mass and up to five spheroids. You will be able to specify the dimensions and mass of each spheroid, to say whether that mass is visible, and to specify whether the spheroid is homogeneous or non-homogeneous.

4.1.1 Units

The unit of mass is 10^9 solar masses. The unit of length is one kiloparsec (kpc). The unit of time is 10^6 years. The constant of gravitation in these units is, numerically, $G = 0.00449897$.

4.1.2 Method of Computation

To explain the program it is necessary to summarize formulas for the gravitational attraction of spheroids. This summary is based on a discussion by Mihalas,[1] section 12.4. This is based, in turn, on a seminal paper by M.Schmidt.[4] These formulas will only be needed if you want to change the law of mass distribution in the spheroids.

Let the boundary of a spheroid have semi-axes a, a, c. We shall only consider oblate spheroids where $a > c$. The *eccentricity* of the spheroid is

$$e = \sqrt{1 - \frac{c^2}{a^2}}. \tag{4.1}$$

We assume that the internal density is constant over any spheroid having the same eccentricity, e. If such a spheroid has semimajor axis α, then we can write the formula for the density, ρ, as

$$\rho = \rho(\alpha).$$

The radial force in the *x-y* plane at a point distant r from the origin can be written as

$$F_r = -\frac{4\pi\sqrt{1 - e^2}}{e^3} rG \int_0^\gamma \rho\left(\frac{r \sin \beta}{e}\right) \sin^2 \beta \, d\beta, \tag{4.2}$$

where

$$\gamma = \arcsin e, \, r < a, \quad \text{and} \quad \gamma = \arcsin\left(\frac{ae}{r}\right), r > a. \tag{4.3}$$

Homogeneous Spheroid. The density, ρ_0 is constant. Then

$$F_r = -\frac{2\pi\sqrt{1 - e^2}}{e^3} rG\rho(\gamma - \sin \gamma \cos \gamma). \tag{4.4}$$

Non-Homogeneous Spheroid. Here we consider just one model for the density, based on Schmidt's work. Let

$$\rho(\alpha) = \rho_0\left(\frac{a}{\alpha} - \frac{\alpha}{a}\right), \quad \alpha \leq a. \tag{4.5}$$

This ensures that the density is zero on the boundary, and is concentrated at the center. Then

$$F_r = -\frac{4\pi\sqrt{1 - e^2}}{e^3} rG\rho_0\left[\frac{ae}{r} \cos \beta + \frac{r}{ae}\left(\cos \beta - \frac{1}{3} \cos^3 \beta\right)\right]_0^q. \tag{4.6}$$

The mass of the spheroid is

$$M = 4\pi \int_0^a \sqrt{1 - e^2}\rho(\alpha)\alpha^2 \, d\alpha = \pi\sqrt{1 - e^2}a^3\rho_0. \tag{4.7}$$

This and Eq. 4.1 make it possible to find ρ_0 from the input quantities a, c, and M.

In the program the galaxy is viewed as projected into the *x-y* plane. So it is necessary to have a formula for the mass inside a cylindrical shell with radii r and $r + dr$. This will be

$$dM = 4\pi r \, dr \int_0^{\sqrt{1-e^2}\sqrt{a^2-r^2}} \rho\left(\sqrt{r^2 - \frac{z^2}{1 - e^2}}\right) dz. \tag{4.8}$$

Using Eq. 4.5, this becomes

$$dM = 4\pi r \, dr \, \rho_0 c\left(\left(1 - \frac{r^2}{2a^2}\right)\ln\left(\sqrt{\frac{a^2}{r^2} - 1} + \frac{a}{r}\right) - \frac{1}{2} \sqrt{1 - \frac{r^2}{a^2}}\right). \tag{4.9}$$

These are the formulas that you must modify if you experiment with other laws of density.

4.1.3   Suggested Changes to the Program

To speed up the animated rotation you might reduce the total number of stars.

More detail could be included in the galactic model by increasing the number of spheroids, or by changing the formula for the density. But if you make this formula more elaborate, it will probably be best to work with a central mass and just one spheroid.

4.1.4   Running the Program

The menu items concerned with the running of the program are

> **Galaxy**   **Rotate**   **Profile**

Galaxy requests data using two input screens. On the first, the user is prompted for the value of the magnitude of the central mass and the number of spheroids. The following screen requests data for each spheroid: the semimajor axis, the semiminor axis, the mass, and whether that mass is visible. If it is visible, then it will be shown in the form of stars. (If you want to see the effect of a central mass alone, enter one for the number of spheroids, enter the radius of your galaxy for the semimajor axis (and some non-zero value for the semiminor axis), enter zero for its mass, and enter "Y" for its visibility.)

In the program the radius of the visible part of the galaxy is divided into 100 sections; the circular velocity is found for each one. The number of stars for a given radius is proportional to the mass of a cylindrical shell around the radius to the total visible mass.

Rotate shows the animated differential rotation of the visible galaxy. Since the program has been set up to show 2,200 stars, the animation may not be very exciting. For this program, the coloring of the stars depends on the initial position angle, so before the rotation starts, the colors are radial; this enables the differential rotation to be seen clearly. The default time interval between successive screens in the animation is equivalent to 10^6 years.

Profile shows the rotation curve for the galaxy.

4.2   Galactic Kinematics and Oort's Constants

In the simplest model for galactic kinematics it is assumed that orbits are circular and coplanar. We shall refer to the "Sun" as having such an orbit. (Strictly, we should replace the Sun by the *dynamical* "local standard of rest." The phrase "local standard of rest" normally refers to the *kinematical* location, found from statistical work on the motions of nearby stars. The "solar motion" refers to the motion of the Sun relative to this latter point. The two standards of rest do not differ greatly.)

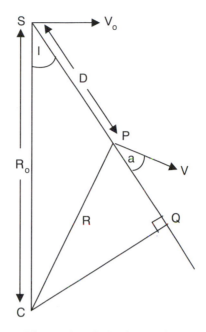

Figure 4.1: Galactic rotation.

4.2.1   Oort's Constants

We need formulas for the radial velocity and proper motion of stars as observed from the Sun. In Figure 4.1, C is the center of the galaxy and S is the Sun, distant R_0 from C. P is a star, distant R from C, and D from the Sun, having galactic longitude l. The velocities of the Sun and star are $V_0 = R_0\omega_0$ and $V = R\omega$, perpendicular to the respective radius vectors $\overrightarrow{CS}$ and $\overrightarrow{CP}$, as shown. Let the velocity of the star make an angle a with SP. Then the radial and tangential velocities as observed from the Sun are

$$V_R = V \cos a - V_0 \sin l = R\omega \cos a - R_0\omega_0 \sin l, \tag{4.10}$$

and

$$V_T = V \sin a - V_0 \cos l = R\omega \sin a - R_0\omega_0 \cos l. \tag{4.11}$$

Now $\angle SPC = \pi/2 + a$, so $\cos a = \sin \angle SPC$. Also, by the law of sines,

$$\frac{\sin \angle SPC}{R_0} = \frac{\sin l}{R}. \tag{4.12}$$

So

$$V_R = R\omega \left(\frac{R_0}{R}\right) \sin l - R_0\omega_0 \sin l$$

$$= R_0(\omega - \omega_0) \sin l. \tag{4.13}$$

To develop the expression for V_T, construct the line CQ, perpendicular to SP, as shown. Then

$$R \sin a = PQ = SQ - D = R_0 \cos l - D. \tag{4.14}$$

So

$$V_T = (R_0 \cos l - D)\omega - R_0\omega_0 \cos l$$

$$= R_0(\omega - \omega_0) \cos l - \omega D . \tag{4.15}$$

These expressions are general, and will be used in the following section on the location of spiral arms. Next, we shall follow Oort's development of approximate expressions, valid for motion in the vicinity of the Sun. Now ω is a function of distance, R, from the galactic center. So, if only first-order terms in $(R - R_0)$ are retained,

$$\omega - \omega_0 = \omega(R) - \omega(R_0) = \frac{d\omega}{dR}\bigg|_{R=R_0} (R - R_0) . \tag{4.16}$$

To the same accuracy,

$$R_0 - R = D \cos l . \tag{4.17}$$

So, since

$$\frac{d\omega}{dR} = \frac{d}{dR}\left(\frac{V}{R}\right) = \frac{1}{R}\frac{dV}{dR} - \frac{V}{R^2} ,$$

then

$$V_R = \left(\frac{V_0}{R_0} - \frac{dV}{dR}\bigg|_{R=R_0}\right)D \sin l \cos l . \tag{4.18}$$

Define *Oort's constant A* by

$$A = \frac{1}{2}\left(\frac{V_0}{R_0} - \frac{dV}{dR}\bigg|_{R=R_0}\right); \tag{4.19}$$

then

$$V_R = AD \sin 2l . \tag{4.20}$$

We have used the trigonometric identity $\sin 2l = 2 \cos l \sin l$.

To find a similar expression for V_T, we can use

$$\omega - \omega_0 = -\left(\frac{V_0}{R_0} - \frac{dV}{dR}\bigg|_{R=R_0}\right)\frac{R - R_0}{R_0}$$

and

$$\omega D = (\omega - \omega_0)D + \omega_0 D \approx \omega_0 D ,$$

correct to the first-order. Then

$$V_T = -\left(\frac{V_0}{R_0} - \frac{dV}{dR}\bigg|_{R=R_0}\right)(R - R_0) \cos l - \omega_0 D$$

$$= \left(\frac{V_0}{R_0} - \frac{dV}{dR}\bigg|_{R=R_0}\right)D \cos^2 l - \frac{V_0}{R_0}D$$

$$= \frac{1}{2}\left(\frac{V_0}{R_0} - \frac{dV}{dR}\bigg|_{R=R_0}\right)D \cos 2l - \frac{1}{2}\left(\frac{V_0}{R_0} + \frac{dV}{dR}\bigg|_{R=R_0}\right)D , \tag{4.21}$$

where we have used the trigonometric identity $2 \cos^2 l = 1 + \cos 2l$. Now define *Oort's constant B* by

$$B = -\frac{1}{2}\left(\frac{V_0}{R_0} + \frac{dV}{dR}\bigg|_{R=R_0}\right)D;\qquad(4.22)$$

then

$$V_T = D(A \cos 2l + B).\qquad(4.23)$$

This can be expressed in terms of the proper motion, μ, since $\mu = V_T/4.74D$:

$$\mu = \frac{A \cos 2l + B}{4.74}.\qquad(4.24)$$

μ is expressed in seconds of arc per year.

A and B are expressed in the units (km/sec)/kpc. Typical values are

$$A = 15(\text{km/sec})/\text{kpc} \quad \text{and} \quad B = -10(\text{km/sec})/\text{kpc}.$$

The program is intended as an illustration to complement instruction on Oort's constants and the extent of their validity in deriving facts about the galaxy. As such, it offers little scope for modification. As remarked earlier, the "Sun" stands, actually, for the local standard of rest (LSR). The routines for plotting radial velocity and proper motion can be modified so that the quantities are observed from a truer Sun that has its solar motion relative to the LSR. This can serve as an introduction to the solar motion.

4.2.2 Running the Program

The menu items concerned with running the program are

      Galaxy   **Location**   **Oort**

Galaxy has two submenus. **Enter data for a galaxy** supervises the building of the galaxy, as before. When the data are complete, the galaxy is plotted. **See the galaxy** plots the galaxy without the need to enter new data.

Location has three submenus. **Choose a local region** prompts the user to click on the Sun and then to drag to define the region. This region must not contain the galactic center nor extend beyond the visible galaxy. (In the program the rotational speed at a given distance from the galactic center is found by interpolation, using the gravitational field of the galaxy; this is only evaluated within the visible part of the galaxy.) This is the local region that is accessible for observations.

The second submenu is **See differential rotation**. The differential rotation, relative to the Sun, of stars in the local region is shown. An arrow points toward the galactic center. This arrow points in a constant direction; so, although the differential rotation relative to the Sun shows up well, the reference system is rotating in space.

The third submenu, **See velocity fields,** shows velocities relative to the Sun in a **fixed** reference frame. There are four options, chosen through the use of the hot keys: the relative velocities, the radial velocities, the tangential velocities, and the proper motions. The last three show well the double sine curves that are a feature of the curves shown in the final menu item.

Oort allows the user to see the radial velocities or proper motions in the local region; the actual observations and the linear approximation are shown, so that you can estimate the validity of the linear approximation. The appropriate values of the Oort constants are shown.

If, when running the program, you want to choose a different local region without entering data for the galaxy, select **See the galaxy** and then **Choose a local region**.

4.3   The Spiral Structure of a Galaxy

The spiral structure of the galaxy can be investigated observationally by looking for the distribution of hydrogen: hydrogen is assumed to be concentrated in the arms. The ground state of atomic hydrogen is split into two levels, depending on whether the spins of the proton and electron are aligned or opposed, the latter having the lower energy. The difference in energy corresponds to a wavelength of 21.1 cm. Transitions, both radiative and collisional, are rare (taking on the order of 10^6 years between collisions), but because of the large number of interstellar hydrogen atoms, the 21 cm line is prominent in galactic radio spectra.

Figure 4.2 shows an imaginary line profile, with the line shifts converted to radial velocities. The intensity is proportional to the number of hydrogen atoms in the line of sight; each peak corresponds to a gas cloud having the appropriate radial velocity. If we assume that the cloud and the Sun are moving in circular orbits in the galactic plane, then, from Eq. 4.13,

$$V_R = R_0((\omega(R) - \omega(R_0)) \sin l,$$

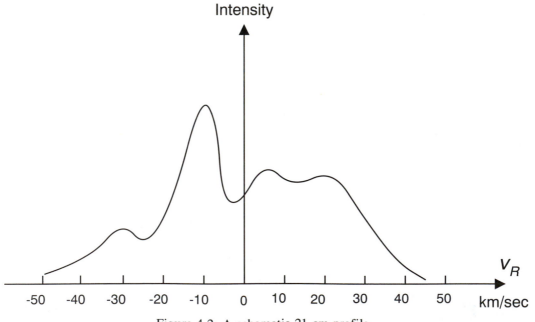

Figure 4.2: A schematic 21-cm profile.

where R is the distance of the cloud from the galactic center. If the rotation curve is known, then R can, in principle, be determined. If the clouds in the line of sight are portions of spiral arms, then, as the galactic longitude of the observations is varied, the positions of the arms can be traced.

This oversimplified discussion ignores the real difficulties encountered in the actual tracing of arms in the galaxy. All the same, it contains the essence of the method, and serves as a good introduction to it. It will be applied in the program for this section.

4.3.1   Running the Program

The menu items involved in running the program are

    Galaxy   **Arms**   **ViewPoint**   **Profiles**

Galaxy oversees, as before, the construction of the galaxy. **Enter data for a galaxy** functions as before. **Plot the galaxy using current data** will plot the galaxy but not any arms. Use it when you want to keep the galaxy, but change the arms.

To make an arm:

1. Select **Arms** and the submenu **Draw a spiral arm**.

2. With the mouse, move the tip of the arrow to the starting point of the arm, and click on it.

3. Move the next point, and click again; a short line joining the points will appear.

4. Continue with this operation. You are allowed up to 50 clicks. You will be informed if you try to exceed this number. You will also be informed if you try to put part of an arm outside the visible galaxy.

5. Double-click at a point to end an arm. (Or it will automatically end at 50 clicks.)

You are allowed up to five arms.

At the start of the program, you will see a galaxy complete with two good-looking spiral arms. You can select these arms later (provided you have three or fewer already drawn) by selecting **Draw the default arms**.

To locate the Sun, select **ViewPoint**, and the submenu **Click on position of viewer**. Then use the mouse to locate the position, and click on it. The rays that appear correspond to the lines of sight of the observations in their default directions. The choice of regular increments of longitude is made in order to simulate a systematic program of observation. However, the operator will also be free to select any longitude, and to see the profile in that corresponding direction.

If you have already selected a viewpoint and wish to change it, then select **Select new position of viewer**.

The program will not accept the selection of **Profile** until at least one arm has been constructed, and the Sun has been located. Once selected, it will show the line profile for galactic longitude of 15° and also a plot of radial velocity as a function of distance in the direction of the observation. You will also see the menu offering the following choices:

Next longitude increases the current longitude by 15°.

Previous longitude decreases the current longitude by 15°.

Choose a longitude starts a dialog where you can move a slider to select a longitude; you will also see an inset plot of the arms and the direction that you are choosing.

Return to profiles returns you from the preceding operation and draws the profile.

Return to main menu does just that.

An emission line is not sharp but is broadened due to the scatter of velocities in the gas around the mean value. In the program each line is plotted as a bell-shaped curve with r.m.s. value "H." This is a parameter in the Procedure Plot21cmLine. Its current value is 5. This can be varied to make the lines more or less sharp.

4.4   Exercises

The Rotation Curve of a Galaxy

4.1   **All mass visible**
Before running the program, you should be familiar with figures for the rotation curve of several galaxies, including our own. For examples, and for additional references, see references 2 and 3. The principal points to be made are as follows:

    a. The observed motion is not Keplerian; that is, it is not dominated by a central mass distribution.

    b. At the edge of the visible galaxy, the rotation curve is not decreasing. This is consistent with the presence of unseen mass that is influencing the kinematics.

I suggest that you start with a central mass only, observing a Keplerian rotation curve. Increase the value of this mass and observe the increased "business" of the rotation close to the center. Then add visible spheroids.

4.2   **Some mass invisible**
Now add a large spheroid of invisible matter and observe the effects.

4.3   **Matching rotation curves**
Find some actual velocity curves and see if you can match them.

4.4 **Motion relative to the Sun**

The local motion, or motion relative to the Sun, is observed in a rotating reference system; such motion is not intuitive. Observe this motion in the program and get used to the reasons for the appearance of plots of radial velocity or proper motion as functions of galactic longitude.

4.5 **Local linearization and Oort's constants**

The determination of Oort's constants leads to estimates of V_0/R_0 and $dV/dR|_{R=R_0}$. Experiment with different local regions. How accurate are these estimates?

4.6 **Galactic structure and Oort's constants: Point mass.**

Suppose that an attempt were to be made to model a galaxy using a central mass only. Could a determination of the Oort's constants repudiate this model?

4.7 **Galactic structure and Oort's constants: Spheroid**

Repeat the preceding exercise, but this time consider a homogeneous spheroid as a model.

4.8 **Radial velocities observed from the Sun**

Suppose that the angular velocity, $\omega(R)$, at a distance R from the galactic center is a monotonically decreasing function of R. Pick a galactic longitude, and consider radial velocities measured in this direction. Show that if $-\pi/2 < l < \pi/2$, the maximum radial velocity will be at the point Q (see Fig. 4.2) where the circular motion is tangent to the line of sight.

4.9 **Distance to the galactic center**

We continue the preceding exercise. Let $V_{max}(l)$ be the maximum radial velocity for given galactic longitude l. Show that

$$V_{max}(l) = V(R_0 \sin l) - V_0 \sin l.$$

Consider this function when l is close to $\pi/2$. Now

$$R_0 \sin l = R_0 - R_0(1 - \sin l),$$

where $(1 - \sin l)$ is small. So

$$V_{max}(l) = V(R_0 - R_0(1 - \sin l)) - V_0 \sin l$$

$$= V(R_0) - dV/dR|_{R=R_0} R_0(1 - \sin l) + \ldots - V_0 \sin l$$

$$\approx 2AR_0(1 - \sin l).$$

Discuss how this result might be used to estimate R_0.

4.10 **The Spiral Structure of a Galaxy**

A good way to use a utility such as this is for one person to design the galaxy and arms, and for a second person to receive the observations, together with data for the Sun and the rotation curve. In this case the curves showing the radial velocity in the line of sight might be omitted. The rotation curve from the first of these projects could be substituted.

References

1. Mihalas, D. *Galactic Astronomy*. San Francisco: W. H. Freeman and Company, 1968.

2. Mihalas, D., Binney, J. *Galactic Astronomy. Structure and Kinematics*. San Francisco: W. H. Freeman and Company, 1981.

3. Rubin, V. C., Ford, W. K., Thonnard, N. Astrophysical Journal Letters. "Extended Rotation Curves of High-Luminosity Spiral Galaxies. IV Systematic Dynamical Properties, $s_a \rightarrow s_c$." **225**:L107, 1978.

4. Schmidt, M. Bulletin Astronomy Institute of the Netherlands "A Model of the Distribution of Mass in the Galactic System" **13**:15, 1956.

5

Interior Model of a Star

Richard T. Kouzes

> Can you bind the Pleiades,
> Can you loose the cord of Orion,
> Can you bring forth the Constellations in their Seasons,
> Or lead out the Bear with its cubs.
> Do you know the Laws of the Heavens...

—Job 38: 31–33.

5.1 Introduction

Understanding the structure of the stars has been one of the main endeavors of astrophysics. A star is a self-gravitating mass of hot gas supported by the nuclear furnace at its center. This program treats a static star in hydrodynamic and thermal equilibrium. This simplified model of a true star gives us insight into the physics which governs stars and their evolution. The model star is a zero-age main sequence (ZAMS) star, meaning that it has a uniform composition of hydrogen, helium, and heavy elements throughout. This model will help you understand the physical processes that exist in stars and should give you some intuitive understanding of how a star's density, temperature, and luminosity depend on its mass.

There is a lot of detailed physics in a star which is simplified in models like this one. Even though the model is simplified, it is a very good approximation to true stars as we understand them and gives global properties of the stellar interior representative of more complete models. While this code is not at the level of sophistication of a research quality stellar model code, it uses the same approach and produces the same result to the limit of the assumptions included, such as the approximation to opacity and convection. Textbooks on stellar models which are useful references include Clayton[3], Collins[4], Kippenhahn and Weigert[7].

One of the current topics of great interest in nuclear and particle astrophysics is the "solar neutrino problem," referring to the deficiency in the observed versus predicted number of neutrinos coming from the Sun. If the Sun is currently generating energy by nuclear fusion, which we believe it is, then the neutrinos should be evident. The model presented here is not complete enough to compute the neutrino spectrum, lacking the stellar evolution from a zero-age star which represents the present Sun, as well as the details of the nuclear reaction processes. But it is a start toward such a model.

The Stellar Evolution (EVOLVE) code, described in the next chapter, provides for evolution of a star through the nuclear burning process.

5.2   The Stellar Interior

5.2.1   Stellar Conditions Assumed

This model of a star assumes a number of simplifications about the stellar condition. A spherically symmetric, non-rotating, self-gravitating model of a star is assumed to be in thermal equilibrium and hydrostatic equilibrium, with a uniform chemical composition throughout. The star is characterized by its mass (M), luminosity (L), effective surface temperature (T_e), radius (R), central density (ρ_c), and central temperature (T_c). Only four of these characteristics can be directly observed, with the central density and central temperature being hidden from observation. For a given mass star, four of these variables are independent and are governed by four differential equations and four boundary conditions. The assumptions of thermal equilibrium plus hydrostatic equilibrium imply that there is no time dependence in the equations describing the star.

5.2.2   Equations Governing the Stellar Interior

There are four differential equations which describe the stellar interior for a static star (e.g., see Kippenhahn and Weigert[7]). Rather than using radius as the independent variable, it is traditional to use mass. The mass enclosed at radius r is given by

$$M_r \equiv \int_0^r 4\pi r^2 \rho \, dr.\qquad(5.1)$$

This definition is used to change radial derivatives to mass derivatives. The first differential equation simply defines the relation between radius (r), mass (M_r), and density (ρ):

$$\frac{dr}{dM_r} = \frac{1}{4\pi r^2 \rho}.\qquad(5.2)$$

The second differential equation arises from the hydrostatic condition that the outward pressure (p) on a shell of matter balances the inward gravitational force everywhere in the star, with G being the gravitational constant:

$$F = 4\pi r^2 \, dp = -\frac{GM_r dM_r}{r^2},\qquad(5.3)$$

or, rearranging,

$$\frac{dp}{dM_r} = -\frac{GM_r}{4\pi r^4}. \tag{5.4}$$

The third equation is a statement of energy conservation. The stellar luminosity at the surface of the star (L_R) represents the power radiated, with no integrated energy at the star's center, $L = 0$ at $r = 0$ and $L = L_R$ at $r = R$. This must be balanced by the fusion energy generated, represented by ϵ, the energy generated per unit mass per second:

$$\frac{dL}{dM_r} = \epsilon. \tag{5.5}$$

The fourth equation is a statement about energy transport. Energy generated in the stellar interior is transported to the surface of the star by radiation, conduction, and convection. The mode of transport depends on the mean free path for photon and particle motion, and on the temperature gradient. Radiation and conduction give rise naturally to equations of the same form. Convection is approximated here by an equation of the same form which is valid in the deep interior of the star, but which ignores the more complicated convection near the stellar surface. The fourth equation describes the relation between the temperature gradient and the energy transport (using the notation of Kippenhahn and Weigert[7]):

$$\frac{dT}{dM_r} = -\frac{GM_rT}{4\pi r^4 p}\nabla, \tag{5.6}$$

where

$$\nabla \equiv \frac{d \ln(T)}{d \ln(p)}. \tag{5.7}$$

If the energy transport is due to radiation and conduction, then

$$\nabla = \nabla_{rad} = \frac{3}{16\pi acG}\frac{\kappa Lp}{M_rT^4}, \tag{5.8}$$

where $a = 7.57 \times 10^{-15}$ erg cm^{-3} K^{-4} is the radiation density constant, and c is the speed of light. The total opacity κ will be described below. If energy transport is by convection, then we approximate ∇ with $\nabla_{adiabatic}$. The two gradients cross rapidly as a function of temperature, and the model simply takes the smaller of the adiabatic or radiative temperature gradient. This is a simplifying approximation to the more complicated reality of a star.

   These four equations, combined with four boundary conditions and four initial guesses, are integrated to yield the solution to the stellar condition. It is important to realize that once the mass of a star and its composition are set, there is a unique solution for the stellar condition. The convergence of the model is very dependent on these initial guesses, and some tables are given later of some reasonable initial guesses for various stellar masses.

5.2.3 Boundary Conditions on the Star

The stellar model is complicated by the fact that the boundary conditions are split between the surface of the star and the center of the star. At the center of the star, we have the conditions that the luminosity is zero (no point energy source at the center) and the mass is zero:

$$L = 0 \text{ and } M_r = 0 \text{ at } r = 0. \tag{5.9}$$

The surface of the star is somewhat complicated to define since it is diffuse, but is taken to be the point at which the total mass is reached, the density falls to almost zero, and temperature is related to the luminosity through the black body radiation equation. At the surface then,

$$M_r = M \text{ and } L = 4\pi R^2 \sigma T_e^4 \text{ at } r = R, \tag{5.10}$$

where the Stefan-Boltzmann constant is $\sigma = 5.67 \times 10^{-5}$ erg cm^{-2} s^{-1} K^{-4}. The Eddington approximation to the stellar atmosphere is made where the optical depth is taken as zero with $\rho = 10^{-12}$ g/cc. The "effective temperature," or photosphere temperature, is related to the "surface temperature" T_s by the definition

$$T_e \equiv 2^{1/4}T_s. \tag{5.11}$$

5.2.4   Details of Stellar Structure

Chemical Composition

The chemical composition of our model star is represented by the three variables X, Y, and Z, where $X + Y + Z = 1$. X represents the mass fraction of hydrogen in the star ($X = 71\%$ for the surface of the Sun), while Y represents the mass fraction of helium in the star ($Y = 27\%$ for the surface of the Sun). Z represents the mass fraction of all heavier elements, often called the "metalicity," in the star ($Z = 2\%$ for the surface of the Sun). These values affect the stellar opacity, nuclear energy generation, and the equation of state.

The chemical composition is assumed to be uniform throughout the star, a condition approximated in zero-age main sequence stars. As a star ages, fusion and convection change the chemical distribution. For example, models give the present central values for the Sun to be $X = 34\%$, $Y = 64\%$ and $Z = 2\%$.

Equation of State

To integrate the four differential equations, the thermodynamic condition at each point in the star must be computed. The quantities needed are the pressure, temperature, radius, and luminosity gradients with respect to mass, as defined by Eqs. 5.2 to 5.6 above (refer to Kippenhahn and Weigert[7] or Stein and Cameron[11]. All equations are computed in the program in units of M_{sun} and R_{sun}.

Equation 5.2 for dr/dM_r is computed from the local values of ρ and r. Equation 5.4 for dp/dM_r is computed from M_r and r. This is then used to compute $d\rho/dM_r$ from the equation

$$\frac{d\rho}{dM_r} = \frac{\rho}{p} \frac{d \ln(\rho)}{d \ln(p)} \frac{dp}{dM_r}, \tag{5.12}$$

where the pressure p is the total pressure. Equation 5.6 requires a computation of the total pressure. The total pressure (p) consists of the radiation (p_{rad}), ion (p_{ion}), and electron pressures (p_e):

$$p = p_{rad} + p_{ion} + p_e. \tag{5.13}$$

The radiation pressure (with radiation constant $a = 7.57 \times 10^{-15}$ erg cm^{-3} K^{-4}) is

$$p_{rad} = \frac{1}{3} aT^4. \tag{5.14}$$

The ion pressure is

$$p_{ion} = \frac{k}{m_p} \frac{\rho T}{\mu_{ion}}, \tag{5.15}$$

where k is the Boltzmann constant, m_p is the proton mass, and μ_{ion} is the mean molecular weight of the ions only (with oxygen representing heavy elements):

$$\mu_{ion} = \frac{1}{X + Y/4 + Z/16}. \tag{5.16}$$

The electron pressure consists of—

the non-degenerate electron pressure,

$$p_{e(non-deg)} = \frac{k}{m_p} \frac{\rho T}{\mu}, \tag{5.17}$$

where the mean molecular weight per free electron is

$$\mu = \frac{2}{1 + X};$$

the non-relativistic degenerate electron pressure,

$$p_{e(deg,non-rel)} = 9.91 \times 10^{12} (\rho/\mu)^{5/3}; \tag{5.18}$$

and the relativistic degenerate electron pressure,

$$p_{e(deg,rel)} = 1.231 \times 10^{15} (\rho/\mu)^{4/3}. \tag{5.19}$$

The total degenerate electron pressure is computed in a way which favors the smaller of the two components

$$p_{e(deg)} = \left[\frac{1}{p_{e(deg,non-rel)}^2} + \frac{1}{p_{e(deg,rel)}^2} \right]^{-1/2}. \tag{5.20}$$

The total electron pressure is then approximately

$$p_e = [p_{e(non-deg)}^2 + p_{e(deg)}^2]^{1/2}. \tag{5.21}$$

Equation 5.4 for dp/dM_r is computed from M_r and r, and then used to compute $d\rho/dM_r$ from Eq. 5.12 using the total pressure to calculate the pressure derivative. Equation 5.5 for dL/dM_r is simply the nuclear power generation per mass ϵ, as described later. Equation 5.6 for dT/dM_r is computed from the temperature gradient, which is computed as the smaller of the radiative or adiabatic gradients:

$$\frac{d \ln(T)}{d \ln(p)}\bigg|_{rad} = \frac{1}{16\pi acG} \frac{\kappa pL}{p_{rad}M}$$

$$\frac{d \ln(T)}{d \ln(p)}\bigg|_{adiabatic} = \left[\frac{\partial \ln(p)}{\partial \ln(\rho)}\bigg|_s \frac{\partial \ln(\rho)}{\partial \ln(T)}\bigg|_s + \frac{\partial \ln(p)}{\partial \ln(T)}\bigg|_\rho \right]^{-1}. \tag{5.22}$$

Opacity

The opacity κ, or cross section per unit mass, is a measure of the mean free path (ℓ) for radiation or conduction under stellar conditions:

$$\ell = \frac{1}{\kappa\rho}. \tag{5.23}$$

The larger the opacity, the shorter the mean free path, and the greater the inhibition to heat transfer. The radiative opacity (κ_{rad}) consists of a number of terms: electron scattering opacity, the "Kramer's term" opacity (i.e., free-free, bound-free, and bound-bound transitions), negative hydrogen ion opacity, and molecular opacity. In addition, electron conductivity produces an opacity (κ_{cond}). In our Sun, the transport of energy is dominated in the central regions by photon radiation, with electron conduction contributing somewhat in the innermost regions, and convection dominating at the surface. In this program, the following equations are used to compute the total opacity.

Electron scattering opacity:

$$\kappa_e = 0.2(1 + X)[1 + 2.7 \times 10^{11}\rho/T^2]^{-1}[1 + (T/4.5 \times 10^8)^{0.86}]^{-1}. \tag{5.24}$$

Kramer's term opacity

$$\kappa_K = 4 \times 10^{25}(1 + X)(Z + 0.001)\rho T^{-3.5}/f_K, \tag{5.25}$$

where $f_K = 2$ is an ad hoc factor modifying the opacity equations used here to improve the model results.

Negative hydrogen ion opacity (for $T < 4 \times 10^4$, otherwise no contribution):

$$\kappa_{H-} = 1.1 \times 10^{-25}(Z\rho)^{0.5}T^{7.7}. \tag{5.26}$$

Molecular opacity (for $T < 4 \times 10^4$, otherwise no contribution):

$$\kappa_m = 0.1Z. \tag{5.27}$$

Conductive opacity (for $\rho > 10^{-5}$, otherwise no contribution):

$$\kappa_{cond} = 2.6 \times 10^{-7}Z_{ave}(T/\rho)^2[1 + (\rho/2 \times 10^6)^{2/3}], \tag{5.28}$$

where $Z_{ave} = (X + 2Y + 8Z)$ is the average nuclear charge (taking oxygen to represent heavy elements).

These opacity components, which have varying frequency dependencies, are combined in a way to give an approximate mean value—

$$\kappa_{rad} = \kappa_m + \left[\frac{1}{\kappa_{H-}} + \frac{1}{\kappa_e + \kappa_K}\right]^{-1} \tag{5.29}$$

and

$$\kappa = \left[\frac{1}{\kappa_{rad}} + \frac{1}{\kappa_{cond}}\right]^{-1}. \tag{5.30}$$

This total opacity is used in the evaluation of the energy transport equation and is evaluated on each step of the integration process in the model.

The opacity used here is an approximation to the true value. The code could be improved by using opacity tables, taking into account partial ionization, and including a calculation of the mixing length to handle convection correctly.

Energy Generation

Energy is generated inside stars through nuclear fusion, the carbon-nitrogen-oxygen (CNO) cycle being first described in detail by Bethe[2] in 1939, and the proton-proton (pp) cycle in 1952 by Saltpeter[9]. The fusion process at stellar temperatures occurs through quantum mechanical barrier penetration, which fortunately occurs at a very slow rate. Only the pp and CNO cycles are included in the model, and only if the temperature is over a million degrees Kelvin. The pp cycle fuses four protons into one helium nucleus, releasing 26.7 MeV of energy (0.71% of the mass), while the CNO cycle performs the same fusion process catalytically releasing 25.0 MeV. The fusion rates can be theoretically derived, and must include a "screening factor" which accounts for the shielding of the nuclear charge by free electrons. The pp and CNO energy generation rates per unit mass (in units of erg s^{-1} g^{-1}) used are from Kippenhahn and Weigert[7]:

$$\mathcal{P}_{pp} = X^2 \rho T_6^{-1/3} g_{11} e^{(14.68 - 33.80 T_6^{-1/3})}, \tag{5.31}$$

where $T_6 = T/10^6$ and the screening factor g_{11} is

$$g_{11} = 1 + 0.0123 T_6^{1/3} + 0.0109 T_6^{2/3} + 0.0009 T_6 \tag{5.32}$$

$$\mathcal{P}_{CNO} = X Z_{CNO} \rho T_6^{-2/3} g_{14} e^{64.33 - 152.28 T_6^{-1/3}}, \tag{5.33}$$

where Z_{CNO} is the mass fraction of CNO elements (taken as $Z/3$ in the model) and where the screening factor g_{14} is

$$g_{14} = 1 + 0.0027 T_6^{1/3} - 0.00778 T_6^{2/3} - 0.000149 T_6 \tag{5.34}$$

The energy generation rate ($\mathcal{P}$) is the sum of $\mathcal{P}_{pp}$ and $\mathcal{P}_{CNO}$, and is evaluated on each step of the integration process in the model. The helium burning rate (3 α's to ^{12}C, yielding 7.3 MeV),

$$\mathcal{P}_{3\alpha} = 5.09 \times 10^{11} \rho^2 Y^3 T_8^{-3} 10^{-18.9/T_8}, \tag{5.35}$$

where $T_8 = T/10^8$, is not computed in this model since it is not normally important until all the hydrogen fuel is consumed, a process important after the end of a star's main sequence life.

The hydrogen depletion rate (in units of g s^{-1} g^{-1}) is also computed as the energy generation rate times the hydrogen-helium mass conversion factor:

$$\frac{dX}{dt} = -1.7 \times 10^{-19} \mathcal{P}. \tag{5.36}$$

The helium depletion rate by conversion into heavy elements (in units of g s^{-1} g^{-1}), which is not included in the model, has a coefficient an order of magnitude larger:

$$\frac{dZ}{dt} = 1.7 \times 10^{-18} \mathcal{P}_{3\alpha}. \tag{5.37}$$

Convection

Convection transports energy through "blobs" of matter being physically transported by the temperature gradient. The "mixing-length" theory for convective transport is not included in the present model. The temperature gradient is, however, calculated from the smaller of the radiative or adiabatic temperature gradients, which accounts partially in the interior for convective energy transport. Near the stellar surface, the temperature gradient makes convection an efficient heat transport process, but this convection is not included in the program.

Degeneracy

Electron, ion, and neutron degeneracy can provide part of the pressure that supports a star. This pressure results from the Pauli exclusion principle preventing any two fermions from occupying the same state. Degeneracy pressure becomes important at high density and low temperature. At very high density, relativistic effects come into play. The program includes the pressure due to a partially degenerate and partially relativistic electron gas.

Neutrinos

Neutrinos are generated by the fusion process, with the overall proton-proton reaction fusing four protons into helium and yielding two neutrinos and 26.7 MeV (4.2×10^{-5} erg) of energy. This is 7×10^{18} erg per gram of hydrogen converted to helium. Thus, one neutrino is produced per 2×10^{-5} erg released in a star by the pp process. The program computes the neutrino release from this process by simply dividing the luminosity by this factor. In the sun, neutrinos carry away about 2% of the energy and the pp neutrinos account for 90% of the neutrinos produced. No attempt is made in this model to compute the neutrino energy spectrum from any fusion processes, but computing the shape of the spectrum is straightforward kinematics. Finding the flux of neutrinos arising from ^7Be and ^8B requires a detailed, very temperature dependent (the flux is proportional to T^{18} [1]) calculation which could be a possible extension to this model.

5.3   Computational Approach

5.3.1   The Present Model

Early computational work on stellar models was carried out by Martin Schwarzchild[10], while L. G. Henyey[6] is associated with the numerical procedures now used in stellar modeling. A "shooting method" is used to integrate the four differential equations (for T, ρ, r and L) from the center of the star outward and from the surface of the star inward to a "fitting point." The fitting point is $M/2$ by default, although it can be varied. This integration is repeated with a variation in each of the four boundary parameters—central temperature, central density, effective temperature, and surface luminosity ($T_{central}$, $\rho_{central}$, $T_{effective}$, and L_R). The deviation in the four values (T, ρ, r, L) at the fitting point, with respect to variation in the boundary parameters, gives corrected values of the four boundary parameters. The user can control the parameter variation to steer the model toward convergence, as described later. The iteration is repeated until the model converges.

5.3.2 Program Procedures

The Procedure **StellarModel** calls Procedures **Outward** and **Inward** to perform the integration from the center out and from the surface in, respectively. After it performs the integrations with variations in each of the boundary parameters, a matrix of derivatives is computed. The Procedure **CorrectBoundary** uses this matrix to compute the corrections to the boundary parameters. The Procedure **StellarModel** iterates until convergence is obtained.

Procedures **Inward** and **Outward** integrate the differential equations using the Procedure **IntegralStep** to make a step in the integration. **IntegralStep** uses Procedure **EvaluateEquations** to analyze the physics equations, and then uses a second-order Runge–Kutta method to make a step in the integration. **EvaluateEquations** calls Procedures **EquationOfState, Opacity** and **NuclearPower** to obtain the physics values, and outputs the derivatives of T, ρ, r and L with respect to mass.

The Procedure **CorrectBoundary** finds the corrections to the boundary parameters by inverting a matrix. The five parameters are

$$x_i \equiv T, \rho, r, L, M_r \text{ for } i = 1 - 5. \tag{5.38}$$

The four differential equations can be written as

$$\frac{dx_i}{dx_5} = y_i. \tag{5.39}$$

The four boundary parameters are

$$z_j \equiv T_e, L, T_c, \rho_c \text{ for } j = 1 - 4. \tag{5.40}$$

The integrations produce a set of four differences at the fitting point (which is at about the radius which contains half the star's mass), for variation in each of the four parameters z_j:

$$\Delta(x_i) \equiv x_{i,center}(z_3, z_4) - x_{i,surface}(z_1, z_2). \tag{5.41}$$

We desire that these differences $\Delta(x_i) = 0$, indicating convergence of the model. To make an iteration, deltas in the boundary parameters (δz_j) must be computed for the four boundary parameters by solving the equation

$$\Delta(x_i) = \sum_1^4 C_{ij} \delta z_j = 0, \tag{5.42}$$

where

$$C_{ij} = \frac{\partial \Delta(x_i)}{\partial z_j}. \tag{5.43}$$

This represents four equations with four unknowns δz_j, which are solved by matrix inversion. The boundary parameters are then modified to make the next iteration:

$$z_j = z_j + \delta z_j. \tag{5.44}$$

This iteration continues until the delta for each parameter is smaller than the desired fitting accuracy.

5.4   Exercises

The following exercises are a few of those possible with the model. A number of extensions to the code are described later to expand the model for other investigations of stellar structure. Users of the program should begin with exploration of the menu selections to become familiar with the program's capabilities. To run a model, enter the initial stellar parameters in the **Boundary Conditions** menu, change the composition in the **Composition** submenu, and run the model from the **Compute** menu.

5.1   The Sun

The sun is the most obvious initial model to investigate, and the program defaults for the initial boundary conditions are those for a zero-age sun-like star. Run the model from the **Compute** menu and watch the convergence process. The results will differ from the present sun since, after 5 billion years, the sun is no longer uniform throughout the star in its composition (X, Y, and Z). What do you learn from the temperature versus mass, power versus mass, and density versus mass plots?

5.2   Boundary Condition Effects

Evaluate the effect of the starting values at the boundaries of the star (center and surface). Set the star mass to 3 (in the **Modify Initial Parameters** menu under **Boundary**), and initiate the relaxation of the star in learning mode, which disables autoconvergence of the model (**Learn About Boundary Effects** under **Compute**). The model will be far from convergence. Use the hot keys to adjust the central temperature, luminosity, central density, and surface temperature to guide the program toward convergence. Repeat this exercise with various masses to gain a feeling for the trends in these parameters with star mass, and to understand which parameters are crucial to convergence. The program will beep when it accepts the hot key input (only one hot key is read per loop).

5.3   Mass Versus Luminosity

Run the code for a series of star masses, e.g., 0.5, 1, 2, 4, 8 solar masses, and note the four boundary parameters' central temperature, central density, effective temperature, and luminosity for each. Make a plot of luminosity versus mass. What do you learn from this plot? Use this information to make a prediction of stellar lifetime as a function of mass.

5.4 **Opacity**

The **Opacity** menu allows you to explore the calculation of opacity and its components. How do the calculations for a mass 1 star compare with values from a reference such as Cox and Tabor[5]?

Determine which component of the opacity dominates under different stellar conditions. To do this, make a plot with $\log(T)$ versus $\log(\rho)$. Compute the opacity for different points on the graph and draw lines which roughly divide the regions where each of the following dominate: electron scattering opacity, electron conductivity opacity, radiative opacity, negative hydrogen ion opacity, and molecular opacity.

Since the opacity calculation is an approximation, a better routine could be produced using opacity tables, including partial ionization, and including mixing length theory for convection, as suggested below under Modifications to the Program.

5.5 **Energy Generation**

What percentage of the star is involved in the energy production process? Within what radius is 90% of the energy produced? How does this vary with the star's mass? Use the **Power** menu to evaluate how density and composition affect the energy production rate. Under what conditions does CNO dominate over pp energy production?

5.6 **Neutrino Generation**

What effect does neutrino loss have on the central temperature of the star? How does the central temperature affect the neutrino production? How does the neutrino flux given by this code compare to the results of sophisticated codes like that of Bahcall?[1] Be sure to use a reasonable composition for this calculation.

5.7 **Equation of State**

The **Calculate Equation of State** screen provides the various pressure components and derivatives under various conditions. A plot of $\log(T)$ versus $\log(\rho)$ can be used to define the regions where various pressure components are most important to the total pressure. Draw a line on this plot representing the conditions in the sun as a function of radius.

Under what stellar conditions does degeneracy become important? Is degeneracy important in the sun?

5.8 **H-R Diagram**

Make a Hertzsprung-Russell diagram, a plot of $\log(L)$ versus $\log(T_e)$. The luminosity log scale should range from -3 to $+7$ and the temperature log scale should run (inverted) from 5 to 2. Run the model for a range of masses (0.5 to 500 solar masses) and fill in the points on the H-R diagram. You should find a zero-age main sequence line produced on the plot.

A plot of $\log(L)$ versus $\log(M)$ is also an informative way to look at the model results.

5.9   **High Mass Stars**

Explore the high mass limit of the model. What limits the model in going to higher masses? What is the limiting mass value? Don't try to jump too high in mass all at once; rather, step up to the desired mass (e.g., run 1, 10, 20, 50, ... masses). This will allow the model to converge from a reasonable starting point. Why are very high mass stars not seen in nature?

5.10   **Low Mass Stars**

Explore the low mass limit of the model. What limits the model at the low mass end?

5.11   **Composition Effect**

Explore the effect of composition on the result of the stellar model. As a star ages, the hydrogen content is depleted and helium content increases, though not uniformly in the star. You can approximate this by adjusting the composition to decrease the hydrogen mass fraction and see what effect this has on the luminosity, effective temperature, central density, and central temperature of the star. For a mass 1 star, make a plot of the resulting stellar radius versus hydrogen mass fraction (e.g., for values 0.1 to 1.0).

5.5   Possible Modifications to the Program

There are a number of extensions that can be made to this model to improve and broaden the physics content. The following is a partial list of these possible extensions.

5.5.1   Opacity Plots

In the exercises above, comparative opacity plots were made in addition to those in the program. This could be done by the code as another plotting extension to the **Opacity** menu, or a fourth plot could be added to the **PlotMoreOpacity** routine. To do this, add the initialization of the plot at the end of the **InitializeMoreOpacityPlot** routine, and add the plot call in the **Complications** routine.

5.5.2   Opacity Tables

The model code could be improved by the use of opacity tables (which are quite large) rather than analytic forms for the opacity. These tables are given, for example, in Cox and Tabor.[5] A table lookup and interpolation routine would then replace the **Opacity** procedure.

5.5.3   Helium Burning

Activate the helium burning calculation in the model. The **NuclearPower** computation needs to call the **NuclearPowerAlpha** procedure. Explore how this rate is affected by the stellar conditions. At what temperature and density does helium burning dominate?

5.5.4 Polytropes

Polytropes are a standard simplified model of stellar structure. This code can be modified to compute polytrope models instead of, or in addition to, the more complete model here. The **EvaluateEquations** and **EquationofState** procedures would be changed for such a calculation. Is there an advantage to a polytrope model?

5.5.5 Neutrino Spectrum

The neutrino calculation at present is only the total flux assuming production of pp neutrinos. Correct the code to account for all neutrinos, and include these values as an energy loss mechanism from the stellar center. Compute the full shape of the neutrino spectrum by including the pp, pep, hep, ^7Be, and ^8B neutrinos. This is a difficult task, but is of great interest to the research community in connection with the solar neutrino problem. Refer to Bahcall.[1]

5.5.6 Non-Zero-Age Star

The model can be modified to consider a non-zero-age star by allowing for an inner core of different composition. Such a two-layer model would have an inner composition depleted in hydrogen, changing the burning conditions. The easiest point to change the composition is at the fitting mass point, so that the **Outward** procedure would use the modified composition, while the **Inward** procedure would be unchanged. The composition menu would be changed to include changing the inner composition.

5.6 *Details of the Program*

5.6.1 Running the Program

This section describes the menu options found in the program. The main menu bar is arranged so that the **Boundary** and **Compute** menus contain the main model computation, while the other main menu items provide related information. A menu item **Getting Started** is provided under the **File** menu to help guide your initial usage of the code.

- **File**:

  - **About STELLAR**: A help screen about this program.
  - **About CUPS**: A help screen about the CUPS project.
  - **Getting Started**: This provides a help screen giving instructions for getting started with the program.

- **File Save**: Writes a disk file of the calculation results containing printable text. The file name begins with the letters "ST" and contains the mass value (e.g., ST001.000 for mass 1). The four fitted parameters are included as well as the values of mass, temperature, density, radius, luminosity, opacity, power, and pressure stepping out through the star in mass. This feature is provided so that model data can be used outside this program; thus, no provision is made to read these files.

- **Configuration**: Allows for modification of the screen colors, changing the temp directory, and providing memory usage information.

- **Exit**: Exit the program.

● **Boundary**: This selection provides for the input of the boundary condition parameters for the model which effect the computation.

- **Modify Initial Parameters**: Change the parameters for the model. The mass is a fixed value, while the four values central temperature, central density, luminosity, and surface temperature are initial guesses for the values that will be determined by the model. The closer these initial guesses are to the solution, the faster the model will converge. A table of initial guesses is provided at the end of this chapter to guide you.

- **Change Compositions**: This selection allows for setting the values of the composition X, Y, and Z in the model. The values of X and Y are entered here, and Z is determined from the constraint that the sum $X + Y + Z$ must equal 1.

- **Explain Stellar Condition**: This selection gives a descriptive screen of the stellar conditions assumed by the program.

- **Explain Stellar Equations**: This selection briefly describes the equations used to solve the model.

- **Explain Convection/Degeneracy**: This selection briefly describes the model's limitations with regard to convection and degeneracy.

- **Neutrino Production**: This selection provides the value of the total neutrino flux from the pp cycle in the star, and the flux at a distance of one astronomical unit, based on the star's luminosity as computed by the model.

- **Reinitialize Model**: This selection reinitializes the model back to the starting parameters when the program was initiated.

● **Compute**: This selection provides options used to initiate the computation of the model.

- **Run Model**: Starts the iterative calculation of the model. The program shows four plots on the screen as it iterates: $\log(Density)$ versus $(M_r/Mass)$, $\log(Temperature)$ versus $(M_r/Mass)$, $\log(Luminosity)$ versus $(M_r/Mass)$, and $\log(Radius)$ versus $(M_r/Mass)$. An option under **Modify Compute Parameters** allows you to have these plots versus $(Radius/Radius_{star})$ instead of $(M_r/Mass)$. The program quits iterating if, after 1000 iterations, the model has not converged, but the computation can be continued by simply reselecting **Run Model**. The **ESC** key can be used to abort the iteration *after* the present loop is finished. There are several hot keys active during the convergence which can be used to modify the boundary value guesses used in the outward and inward integrations, allowing the user to help the program to convergence, and to give a feeling for how these parameters affect the integration. The hot keys are only checked *once per loop,* and only the one last pushed will be seen. The program gives a beep when it accepts the hot key. Sometimes a "loop" can take a long time because the integrated value has become small, which forces the code to make many steps. These hot keys are lowercase or uppercase characters:

* 'c': Decrease the present boundary value of the central temperature by 10%.

* 'C': Increase the present boundary value of the central temperature by 10%.

* 'd': Decrease the present boundary value of the central density by 10%.

* 'D': Increase the present boundary value of the central density by 10%.

* 'e': Decrease the present boundary value of the effective temperature by 10%.

* 'E': Increase the present boundary value of the effective temperature by 10%.

* 'l': Decrease the present boundary value of the luminosity by 10%.

* 'L': Increase the present boundary value of the luminosity by 10%.

* '+': Increase by a factor of 2 the percentage used as the delta in the boundary condition changes above (10% by default).

* '-': Decrease by a factor of 2 the percentage used as the delta in the boundary condition changes above.

* 'p' or 'P': Pause the computation (restart with any key).

* 'r' or 'R': Replot the screen.

- **Learn About Boundary Effects**: This option runs the model iteration just as in **Run Model**, except the automatic correction of the boundary starting values (central density, central temperature, effective temperature, and luminosity) is disabled. Use this mode of the program to learn about the effects of changing each boundary value. You will need to use the "+" and "-" hot keys to adjust the percent change in the boundary value on each iteration in order to get close enough to the correct boundary value to meet the convergence criterion. The program shows in a message box the last delta values for each parameter, which are the fractional differences between the inward and outward integrations. You must make these smaller than the **Precision of Fit** value set in the **Modify Compute Parameters** menu item in order to have reached convergence. Explore the effect of each boundary value on the solution generated by the model.

- **Show Numerical Results**: Shows a table of some of the numerical results from the calculation for mass, temperature, density, radius, and luminosity.

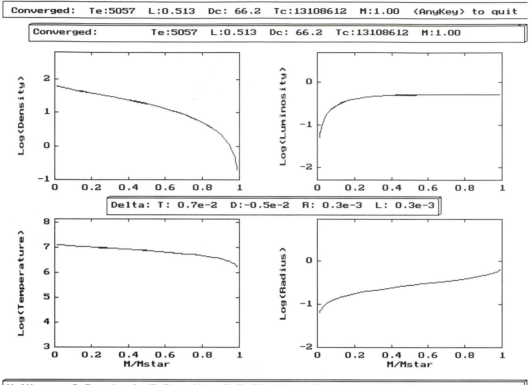

Figure 5.1: The computer screen showing the results of the model computation.

A much more complete set of numerical results is provided in the output file generated in the **File Save** option of the **File** menu.

– **Plot Results**: Plots the results of the last computation, as done during **Run Model**. You can zoom in on a single plot by clicking on it with the mouse. Another click restores the plots. Figure 5.1 shows the results screen.

– **Plot More Results**: Plots more of the results of the last computation, including opacity, power, pressure, and temperature gradient versus (M_r/*Mass*). An option under **Modify Compute Parameters** allows you to have these plots versus (*Radius*/*Radius$_{star}$*) instead of (M_r/*Mass*). You can zoom in on a single plot by clicking on it with the mouse. Another click restores the plots. Figure 5.2 shows this screen.

– **Modify Compute Parameters**: Adjust the fitting criteria. The value of the precision of the fit to the mass at the fitting point may be adjusted, as can the fitting mass point. Sound can be turned on or off here, producing a beep at the end of each iteration through the convergence loop. A toggle option, **Plot Versus Radius Instead of Mass,** allows for the plots under the **Compute** menu to be shown versus (*Radius*/*Radius$_{star}$*) instead of (M_r/*Mass*).

• **Power**: This selection provides a menu of energy generation related items.

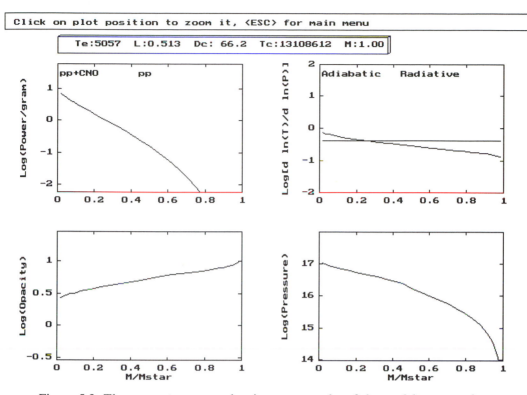

Figure 5.2: The computer screen showing more results of the model computation.

- **Calculate Power**: This selection allows for computation of the fusion energy generation rate under varying conditions of temperature, density, and composition. The pp, CNO, and 3α rates are computed as well as the hydrogen depletion rate.

- **Plot Power Versus Temperature**: A plot is made of power per gram versus temperature using the equations for energy generation given earlier. The present values of density and composition are used in the calculation of the plotted data. Figure 5.3 shows this screen.

- **Eq. State**: This selection provides a menu of equation of state related items.

  - **Calculate Equation of State**: This selection allows for computation of the equation of state parameters under varying conditions of temperature, density, and composition. The pressure components are shown as well as the adiabatic and radiative temperature gradients.

  - **Plot Equation of State**: This selection plots the equation of state parameters versus temperature, including ion pressure, degenerate electron pressure, non-degenerate electron pressure, and total electron pressure. Lines are plotted for the present density and for the present density/10. A slider is provided to adjust the density used in the calculation, and is accepted by a mouse double-click or an **Enter** keystroke.

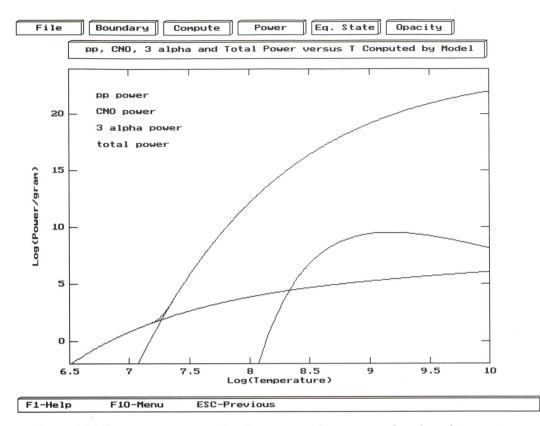

Figure 5.3: The computer screen showing computed power as a function of temperature.

- **Plot More Equation of State**: This selection plots more of the equation of state parameters versus temperature, including radiative pressure, total pressure, adiabatic temperature gradient, and radiative temperature gradient. Lines are plotted for the present density and for the present density/10. A slider is provided to adjust the density used in the calculation, and is accepted by a mouse double-click or an **Enter** keystroke.

- **Opacity**: This selection gives an additional menu of complications to the stellar model having to do with opacity.

  - **Calculate Opacity**: This selection allows for computation of the opacity under various conditions of temperature, density, and composition. Each opacity component is shown, as well as the total.

  - **Plot Opacity**: This selection plots some of the opacity components, including electron, Kramer's, H⁻, and molecular opacities. Lines are plotted for the present density and for the present density/10. A slider is provided to adjust the density used in the calculation, and is accepted by a mouse double-click or an **Enter** keystroke.

  - **Plot More Opacity**: This selection plots some more of the opacity components, including radiative, conductive and total opacities. Lines are plotted for the present density and for the present density/10. A slider is provided to adjust the density used in the calculation, and is accepted by a mouse double-click or an **Enter** keystroke.

5.6.2   Structures of the Program

The program STELLAR consists of one large program, STELLAR.PAS, rather than units, which simplifies maintenance of the code. The following gives the program structure in terms of the menus:

Main Menu

– File—File Save and Quit

* About Stellar—About This Program

* About CUPS—About the CUPS Project

* Getting Started—Guide to Initial Usage of the Program

* File Save—Save Parameters to a Disk File

* Configuration—Change or Examine the Program Environment

* Exit—Exit the Program

– Boundary—Enter the Mass and Four Boundary Initial Conditions

* Modify Parameters—Modify the Model Parameters

* Change Composition—Modify the Stellar Chemical Composition

* Explain Conditions—Describe the Assumed Stellar Conditions

* Explain Equations—Describe the Equations Used

* Explain Convection/Degeneracy—Describe the Effects of Convection and Degeneracy

* Neutrino Production—Compute the Neutrino Production Rate

* Reinitialize—Restore Parameters to Their Initial Values

– Compute—Perform Iterative Computation and Show Graphical Results

* Run Model—Run the Model Calculation

* Learn About Boundary Effects—Run the Model Calculation Without Autoconvergence

* Show Results—Display Model Results

* Plot Results—Plot Model Calculation Results

* Plot More Results—Plot Model Calculation Results

* Modify Compute Params—Modify Model Parameters

– Power—Compute the Nuclear Fusion Energy Generation

  * Calc Power—Compute the Nuclear Fusion Energy Generation

  * Plot Power—Plot the Nuclear Fusion Energy Generation

– Equation of State—Compute the Equation of State Conditions

  * Calc Eq of State—Compute the Equation of State Values

  * Plot Eq of State—Plot the Equation of State Values

  * Plot More Eq of State—Plot the Equation of State Values

– Opacity

  * Calc Opacity—Compute the Opacity Components for Entered Conditions

  * Plot Opacity—Plot the Opacity Components

  * Plot More Opacity—Plot the Opacity Components

5.7 Table of Initial Guesses

This table provides a range of initial guess values for use in the model. These come from a model calculation similar to this one. The units of M, R, and L are in terms of solar mass, solar radius and solar luminosity. Temperatures are in Kelvin, and density is in g/cc.

X	Z	M	T_eff	L	T_central	rho_cen	R
.7	.0300	.10	2418.	.001	4384284.	335.51	.14
.7	.0300	.13	2619.	.001	5046600.	225.89	.17
.7	.0300	.16	2778.	.002	5624695.	166.65	.20
.7	.0300	.20	2923.	.004	6191529.	127.85	.24
.7	.0300	.25	3062.	.006	6774837.	100.09	.28
.7	.0300	.32	3199.	.010	7392628.	79.40	.32
.7	.0300	.40	3338.	.016	8059312.	63.59	.38
.7	.0300	.50	3490.	.025	8685568.	54.03	.44
.7	.0300	.63	3767.	.050	9547666.	56.51	.53
.7	.0300	.79	4330.	.140	11135142.	61.82	.67
.7	.0300	1.00	5316.	.469	13449176.	69.89	.82
.7	.0300	1.26	6778.	1.574	16428565.	77.64	.92
.7	.0300	1.58	8507.	4.861	19168864.	72.09	1.02
.7	.0300	2.00	10378.	14.282	21057072.	58.08	1.18
.7	.0300	2.51	12482.	40.598	22635808.	44.75	1.38
.7	.0300	3.16	14829.	110.255	24132112.	33.92	1.61
.7	.0300	3.98	17382.	283.465	25603584.	25.54	1.87
.7	.0300	5.01	20100.	687.543	27058048.	19.22	2.18
.7	.0300	6.31	22940.	1577.245	28497136.	14.53	2.54

X	Z	M	T_eff	L	T_central	rho_cen	R
.7	.0300	7.94	25864.	3435.575	29908976.	11.09	2.95
.7	.0300	10.00	28860.	7146.598	31282224.	8.58	3.41
.7	.0300	12.59	31901.	14236.375	32613824.	6.75	3.94
.7	.0300	15.85	34970.	27264.621	33900096.	5.42	4.54
.7	.0300	19.95	38019.	50176.324	35131840.	4.42	5.21
.7	.0300	25.12	41020.	88960.812	36299088.	3.67	5.96
.7	.0300	31.62	43903.	152019.437	37410768.	3.10	6.81
.7	.0300	39.81	46612.	250725.875	38467808.	2.65	7.75
.7	.0300	50.12	49125.	400312.187	39454592.	2.29	8.82
.7	.0300	63.10	51404.	620724.625	40382848.	1.99	10.03
.7	.0300	79.43	53432.	936911.375	41257344.	1.75	11.40
.7	.0300	100.00	55220.	1381015.000	42081984.	1.55	12.96
.7	.0300	125.89	56767.	1993418.000	42844576.	1.37	14.74
.7	.0300	158.49	58090.	2825523.000	43581072.	1.22	16.76
.7	.0300	199.53	59197.	3941841.000	44268752.	1.09	19.06
.7	.0300	251.19	60131.	5423734.000	44915808.	.98	21.67
.7	.0300	316.23	60912.	7372234.000	45529952.	.88	24.62
.7	.0300	398.11	61546.	9917409.000	46120784.	.79	27.96
.7	.0300	501.19	62087.	13222101.000	46687232.	.71	31.74
.8	.0001	.13	3223.	.002	4901155.	353.43	.14
.8	.0001	.16	3470.	.004	5618221.	243.11	.18
.8	.0001	.20	3675.	.007	6261784.	181.38	.21
.8	.0001	.25	3861.	.012	6899205.	140.12	.25
.8	.0001	.32	4092.	.021	7471336.	128.44	.29
.8	.0001	.40	4559.	.051	8470304.	133.47	.37
.8	.0001	.50	5376.	.156	10020696.	140.77	.46
.8	.0001	.63	6572.	.484	11997780.	148.25	.54
.8	.0001	.79	8132.	1.400	14348290.	154.24	.60
.8	.0001	1.00	10016.	3.734	17076496.	158.12	.65
.8	.0001	1.26	12181.	9.226	20202256.	159.85	.69
.8	.0001	1.58	14568.	21.311	23686304.	158.09	.73
.8	.0001	2.00	16998.	46.387	27126672.	145.65	.79
.8	.0001	2.51	19266.	96.828	29908976.	120.64	.89
.8	.0001	3.16	21449.	198.061	32151360.	93.89	1.03
.8	.0001	3.98	23654.	400.405	34158656.	71.37	1.20
.8	.0001	5.01	25942.	802.045	36082624.	53.90	1.42
.8	.0001	6.31	28366.	1593.306	37983696.	40.82	1.67
.8	.0001	7.94	30953.	3134.004	39883776.	31.16	1.97
.8	.0001	10.00	33721.	6095.359	41792752.	24.10	2.31
.8	.0001	12.59	36686.	11673.437	43691632.	18.92	2.70
.8	.0001	15.85	39811.	21922.973	45572320.	15.12	3.14
.8	.0001	19.95	43072.	40188.285	47402192.	12.31	3.63
.8	.0001	25.12	46387.	71597.750	49192816.	10.19	4.18
.8	.0001	31.62	49693.	123651.500	50909488.	8.58	4.79
.8	.0001	39.81	52893.	206918.250	52552784.	7.32	5.47
.8	.0001	50.12	55937.	335427.937	54100320.	6.33	6.23
.8	.0001	63.10	58749.	527592.000	55577120.	5.52	7.08

X	Z	M	T_eff	L	T_central	rho_cen	R
.8	.0001	79.43	61305.	807790.688	56963744.	4.86	8.05
.8	.0001	100.00	63577.	1206143.000	58277216.	4.30	9.14
.8	.0001	125.89	65569.	1761160.000	59511040.	3.82	10.38
.8	.0001	158.49	67298.	2521150.000	60673648.	3.42	11.80
.8	.0001	199.53	68754.	3548943.000	61787360.	3.06	13.40
.8	.0001	251.19	70000.	4919248.000	62819744.	2.75	15.23
.8	.0001	316.23	71039.	6731300.000	63811920.	2.48	17.29
.8	.0001	398.11	71912.	9107529.000	64758496.	2.24	19.64
.8	.0001	501.19	72611.	12203837.000	65659952.	2.02	22.28
.8	.0001	630.96	73198.	16210676.000	66527280.	1.83	25.28
.7	.0001	.10	3049.	.001	4544638.	508.86	.12
.7	.0001	.13	3360.	.003	5416243.	316.08	.15
.7	.0001	.16	3581.	.005	6106586.	228.82	.18
.7	.0001	.20	3777.	.008	6765483.	174.46	.21
.7	.0001	.25	3975.	.014	7351884.	146.86	.25
.7	.0001	.32	4358.	.030	8192159.	151.84	.31
.7	.0001	.40	5069.	.089	9616095.	160.62	.39
.7	.0001	.50	6172.	.283	11526535.	170.80	.47
.7	.0001	.63	7661.	.851	13838892.	180.05	.53
.7	.0001	.79	9513.	2.364	16550096.	187.02	.57
.7	.0001	1.00	11673.	6.055	19678912.	191.21	.61
.7	.0001	1.26	14080.	14.395	23184624.	190.99	.64
.7	.0001	1.58	16546.	31.982	26680832.	177.58	.69
.7	.0001	2.00	18854.	67.624	29505088.	147.67	.78
.7	.0001	2.51	21052.	139.508	31761264.	114.95	.90
.7	.0001	3.16	23265.	283.726	33767424.	87.22	1.05
.7	.0001	3.98	25550.	571.083	35686192.	65.71	1.23
.7	.0001	5.01	27964.	1138.675	37583584.	49.58	1.45
.7	.0001	6.31	30535.	2250.604	39481712.	37.71	1.71
.7	.0001	7.94	33297.	4397.426	41399856.	29.00	2.01
.7	.0001	10.00	36241.	8472.266	43311280.	22.63	2.36
.7	.0001	12.59	39373.	16036.113	45217008.	17.98	2.75
.7	.0001	15.85	42648.	29655.090	47086448.	14.53	3.18
.7	.0001	19.95	46015.	53419.461	48909888.	11.97	3.67
.7	.0001	25.12	49397.	93303.750	50663840.	10.02	4.21
.7	.0001	31.62	52711.	157869.937	52347920.	8.51	4.81
.7	.0001	39.81	55873.	258820.750	53951136.	7.33	5.48
.7	.0001	50.12	58816.	411527.375	55462768.	6.37	6.24
.7	.0001	63.10	61518.	635767.937	56898576.	5.59	7.09
.7	.0001	79.43	63929.	957412.062	58250576.	4.94	8.05
.7	.0001	100.00	66069.	1408636.000	59524640.	4.39	9.15
.7	.0001	125.89	67920.	2030479.000	60729232.	3.92	10.39
.7	.0001	158.49	69502.	2874077.000	61858080.	3.51	11.81
.7	.0001	199.53	70843.	4005888.000	62935824.	3.15	13.41
.7	.0001	251.19	71978.	5505520.000	63958176.	2.83	15.23
.7	.0001	316.23	72912.	7478235.000	64922736.	2.56	17.30
.7	.0001	398.11	73688.	10050712.000	65841552.	2.31	19.64
.7	.0001	501.19	74336.	13393681.000	66741792.	2.09	22.28

Acknowledgments

The approach to the computation used here is based on that developed by Bohdan Paczynski of Princeton University and is gratefully acknowledged.[8]

Bibliography

1. Bahcall, J. N. *Neutrino Astrophysics*. Cambridge: Cambridge Universty Press, 1989.

2. Bethe, H. Physical Review. **55:**434, 1939.

3. Clayton, D. *Principles of Stellar Structure and Nucleosynthesis*. Chicago: University of Chicago Press, 1983.

4. Collins II, G.W. *The Fundamentals of Stellar Astrophysics*. New York: W. H. Freeman and Company, 1989.

5. Cox, A. N., Tabor, Y. E. Radiative opacity tables of 40 stellar mixtures. Astrophysics Journal Supplement **31:**271, 1976.

6. Henyey, L. G., Wilets, L., Böhm, K. H., Lelevier, R., and Levee, R. D. "A Method for Automatic Computation of Stellar Evolution," Astrophysics Journal **129:**628, 1959.

7. Kippenhahn, R., Weigert, A. *Stellar Structure and Evolution*. Berlin: Springer-Verlag, 1990.

8. Paczynski, B. Private communication. Princeton, NJ: Princeton University.

9. Salpeter, E. E. Physical Review **88:**547, 1952.

10. Schwarzchild, M. *Structure and Evolution of the Stars*. Princeton, NJ: Princeton University Press, 1958.

11. Stein, R. F., Cameron, A. G.W. *Stellar Evolution*. New York: Plenum Press, 1966.

6

Stellar Evolution

Richard T. Kouzes

> Twinkle, twinkle, little star,
> How I wonder what you are,
> Up above the world so high,
> Like a diamond in the sky.
>
> —Jane Taylor (1806)

6.1 Introduction

Stars are not static objects, but rather are dynamic, evolving, self-gravitating masses of hot gas supported by the nuclear furnace at their center. Stars begin as contracting clouds of dust and gas and evolve from protostars through a contraction phase leading to nuclear ignition and onto the main sequence, where they spend the majority of their lifetime fusing hydrogen into helium. When the hydrogen fuel is depleted, the star evolves rapidly off the main sequence to its eventual demise as a white dwarf or a supernova. Understanding the structure of the stars has been one of the main endeavors of astrophysics, which has tried to answer questions such as, How do stars evolve onto the main sequence? What is the physical basis of the main sequence? What powers the star for billions of years? Why do they evolve off the main sequence? What determines a star's lifetime? Why are stars as stable as they are?

This simulation looks at the evolution of a star through its life cycle. The Interior Model of a Star (STELLAR) program discussed in the last chapter is the basis for this Stellar Evolution code and should be well understood before exploring the evolution process. The equations governing the star under evolution are extensions to the static star equations, including the time dependence ignored in the static star model. This evolution model contains simplifications, but this simplified model still gives us insight into the physics which governs stellar evolution.

This program divides the stellar evolution process into two parts: an approximation to the pre-main sequence evolution, and evolution from the main sequence.

Use this model to explore the time dependence of the stellar process and to discover how stars are created, live, and die. It will help give an understanding of the time scales involved in stellar processes and the role stars play in nucleosynthesis.

6.2 The Stellar Interior

The evolving star model is based on the same physics as the static star model, with extensions. The following descriptions will concentrate on the differences in the equations needed for the evolution code versus the static case, and will assume the background presented in the previous chapter.

6.2.1 Stellar Conditions Assumed

This model assumes a number of simplifications about the stellar condition. A spherically symmetric, non-rotating, self-gravitating model star is assumed with an initially uniform chemical composition throughout. The star is characterized by its mass (M), luminosity (L), effective surface temperature (T_e), radius (R), central density (ρ_c), and central temperature (T_c). Additionally, the composition of the star as a function of radius, and as a function of time, must be specified. For a given mass star, these variables are governed by five differential equations. This model is divided into pre-main sequence evolution and evolution starting from the zero-age main sequence (ZAMS).

6.2.2 Equations Governing the Stellar Interior

There are five differential equations which describe the stellar interior for an evolving star (see, for example, Kippenhahn and Weigert[3]). Four of these equations were used in the static star case, with the time dependence removed, and the fifth equation is a new one describing the star's composition change with time. The equations will now be written as partial differential equations since time dependence is explicity included.

The first differential equation defines the relation between radius (r), mass (M_r), and density (ρ), and is the same as in the static case:

$$\frac{\partial r}{\partial M_r} = \frac{1}{4\pi r^2 \rho}. \tag{6.1}$$

The second differential equation gives the acceleration of a mass shell due to the difference in the outward pressure (p) on the shell of matter and the inward gravitational force on the shell everywhere in the star, with G being the gravitational constant:

$$\frac{1}{4\pi r^2}\frac{\partial^2 r}{\partial t^2} = -\frac{\partial p}{\partial M_r} - \frac{GM_r}{4\pi r^4}. \tag{6.2}$$

This equation describes hydrostatic equilibrium if the acceleration vanishes, and the mass shell is at rest or in constant motion. The hydrostatic condition is a very good approximation if the acceleration is small and the configuration moves through neighboring near equilibrium states. This evolution from ZAMS model does just this, finding a new hydrostatic equilibrium after each time step.

The third equation is a statement of energy conservation. The stellar luminosity at the surface of the star (L_R) represents the power radiated. This must be balanced by the fusion energy generated per unit mass per second (represented by ϵ), the energy carried away by neutrinos (ϵ_v) which is presently ignored, and the change in the internal energy:

$$\frac{\partial L}{\partial M_r} = \epsilon - \epsilon_v - c_p \frac{\partial T}{\partial t} + \frac{\delta}{\rho} \frac{\partial p}{\partial t}, \qquad (6.3)$$

where

$$\delta \equiv -\left(\frac{\partial \ln \rho}{\partial \ln T}\right)_p. \qquad (6.4)$$

The fourth equation, which assumes hydrostatic equilibrium, is a statement about energy transport, and is the same as that used in the static star case (using the notation of Kippenhahn and Weigert[3]):

$$\frac{\partial T}{\partial M_r} = -\frac{G M_r T}{4\pi r^4 p} \nabla, \qquad (6.5)$$

where

$$\nabla \equiv \frac{d \ln(T)}{d \ln(p)}. \qquad (6.6)$$

If the energy transport is due to radiation and conduction, then

$$\nabla = \nabla_{rad} = \frac{3}{16\pi a c G} \frac{\kappa L p}{M_r T^4}, \qquad (6.7)$$

where $a = 7.57 \times 10^{-15}$ erg cm^{-3} K^{-4} is the radiation density constant, c is the speed of light, and κ is the total opacity. If energy transport is by convection, then we approximate ∇ with $\nabla_{adiabatic}$, and the model simply takes the smaller of the two temperature gradients, the adiabatic or radiative temperature gradient. This is a reasonable simplifying approximation.

The fifth equation describes the rate of change of composition of the star for each nuclear species X_i, with mass m_i, in terms of the reaction rates (r_{ij}) for production and destruction of that species:

$$\frac{\partial X_i}{\partial t} = \frac{m_i}{\rho}\left(\sum_j r_{ji} - \sum_k r_{ik}\right) \text{ where } i = 1, \ldots I, \qquad (6.8)$$

with the constraint that $\sum_i X_i = 1$.

These five equations, combined with boundary conditions and initial guesses, are integrated to yield the solution to the stellar condition for the evolution in this model starting at the ZAMS. The boundary conditions are the same as for the static star.

6.2.3 Details of Stellar Structure: ZAMS and Beyond

The stellar interior used for the evolution starting at the ZAMS is the same as that used in the previous chapter and will only be briefly described here, emphasizing the changes.

The chemical composition of our model star is represented by the three variables X, Y and Z, where $X + Y + Z = 1$, corresponding to X_i for i = 1, 2, 3. X represents the mass fraction of hydrogen in the star, while Y represents the mass fraction of helium in the star. Z represents the mass fraction of all heavier elements in the star. The star's nitrogen content, which is needed to compute the carbon-nitrogen-oxygen (CNO) cycle energy release, is assumed to be $Z/3$. The chemical composition of the protostar is assumed to be initially uniform throughout the star (a condition approximated in ZAMS stars) until the star reaches the main sequence. As a star ages, fusion and convection change the chemical distribution. The evolution model divides the star into layers (50) of equal mass. After an equilibrium state is determined by relaxation, the composition of each layer is modified as a result of any fusion power generation in it. Convection of material between layers is not presently performed. The time step used by the model to change the composition after each evolution step is set to keep the largest composition change in any layer to less than a factor (2% by default). This gives large time steps (of order one billion years) for young stars, and smaller time steps as the star moves into shell burning.

To integrate the differential equations, the thermodynamic condition at each point in the star must be computed. The quantities needed are the pressure, temperature, radius, and luminosity gradients with respect to mass, as defined by Eqs. 6.1 to 6.8 above. All equations are computed in the program in units of M_{sun} and R_{sun}. This calculation is the same as in the static star case.

Energy is generated inside of stars through nuclear fusion, which occurs through barrier penetration at stellar temperatures. The proton-proton (pp) and CNO cycles are included in the model, where the temperature is over a million degrees Kelvin. The three helium (3α) fusion process is included in the model and contributes above 10^8 Kelvin. The pp, CNO, and 3α energy generation rates per unit mass used are from Kippenhahn and Weigert.[3]

6.3 Stellar Evolution

The details of stellar evolution are governed by the mass of the star. A body with a mass of greater than about 100 solar masses will blow itself apart before forming because the radiation pressure exceeds the gravitational attraction. An object with a mass less than about 8% of a solar mass will never reach a central temperature high enough to trigger hydrogen fusion and will not form a star. Stars form then in the mass region between about 0.08 and 100 solar masses.

6.3.1 Stages of Evolution

A star begins its path toward existence as a large, tenuous, optically invisible cloud of gas and dust, such as those found in the disks of spiral galaxies. These clouds condense under their own gravitational attraction, triggered by a density fluctuation in the cloud, initially forming a slightly denser nucleus, which then accretes material from the cloud. The cloud collapses due to gravitational attraction, under the constraint of the laws of thermodynamics, through a series of near equilibrium states. The cloud collapses from the inside out, with material at the center free-falling first and the outer material remaining stationary. The collapse spreads outward through the cloud. The rate of mass accretion onto the core depends only on the cloud's initial temperature—the higher the temperature, the faster the accretion rate. It takes from 100,000 to 1 million years for a solar mass to accumulate at the core of the cloud. The gravitational potential energy of the in-falling material is converted to thermal energy in the nucleus, forming a shock front at the surface where the matter is slowed by viscous forces. This nucleus forming at the center of the cloud is a "protostar." Star models usually treat the protostar separately from the cloud, modeling it as a star with accretion. The shock front at the protostar surface heats the gas to about a million degrees, which then rapidly cools by radiation. This radiation is obscured from our view by absorption in the surrounding cloud. (As a result, there is no definitive optical telescope evidence of protostar formation). This radiation also slows the accretion rate of material onto the protostar.

As the protostar gains mass and collapses, the interior temperature rises. Pressure forces build, with motion being converted to a temperature rise in the protostar. The protostar collapse is a slow quasistatic one. The virial theorem becomes a good approximation, with half of the gravitational energy being converted to kinetic energy, and the other half being radiated. The radiation escapes at first, but as the pressure and density build, the protostar becomes opaque, and thermodynamic equilibrium is established for the first time, with the radiation field being described by a local temperature related to the kinetic energy of the gas. Some radiation escapes, with much of it being consumed in ionizing the cloud. Hydrostatic equilibrium is established when the interior becomes largely ionized. For the sun, this occurred when it had shrunk to only about 60 times its present size.

As the star approaches its final luminosity, the evolution slows and a core in radiative equilibrium grows. The core grows and the convective zone shrinks, with the central temperature increasing. Once the protostar mass grows to a few tenths of a solar mass, fusion of deuterium begins and plays a major role in the protostar evolution, despite the rarity of this isotope. The deuterium burns in a shell on the protostar, and results in a wind which blows off material and halts the accretion process. The protostar collapse eventually causes the interior temperature to reach 10 million degrees Kelvin, a value where hydrogen fusion begins. The star rapidly equilibrates. The energy generated equals that radiated, equilibrium is established, and the star has reached the zero-age main sequence.

When all the hydrogen fuel in the core is consumed, hydrogen continues to burn in a shell around an isothermal core as a red giant star, until the temperature rises high enough for helium burning. The star then evolves rapidly through helium and heavier element burning until, with its fuel consumed, it becomes a supernova or a white dwarf.

Consider the evolution process for the Sun, a star of one solar mass. A protostar forms in a cloud and accretes material onto its surface. It has a larger radius and luminosity, but a lower surface temperature, than it will eventually have on the main sequence. The protostar has a high opacity due to its low temperature,

and convection transports the energy being converted from gravitational potential energy to the surface, where it is radiated away. This efficient radiation process forces the star to contract, decreasing its surface area, and thus its radiation
rate. The core heats up, reaching a few million degrees Kelvin, triggering hydrogen fusion, and the star moves into hydrostatic and thermal equilibrium as fusion
takes over the energy generation from gravity. The Sun has taken 50 million
years to reach the zero-age main sequence, where it will remain in its hydrogen
burning phase for about 10 billion years. A heavier star would reach the main sequence with a higher luminosity, consuming its fuel much more rapidly, and would
last a much shorter time in the main sequence (only 12 million years for a 15 solar mass star).

When the Sun has consumed most of the hydrogen in its core, its core temperature rises slowly, and it contracts, while its luminosity increases. Hydrogen
burning continues in a shell around the core, which continues to contract under its
gravitational attraction. The hydrogen shell increases in temperature, speeding
up the reaction and heating the outer part of the star. The radius of the star thus
increases and its surface temperature decreases, increasing the opacity and forc-
ing convection to transport energy to the surface. The Sun has now become a red
giant, with a lower temperature but a higher luminosity

The core has now reached a high enough density that it has become a degenerate electron gas, providing the pressure to stop its contraction, a pressure which
is independent of temperature as long as the gas remains degenerate. The core
temperature rises slowly, driven by the shell burning, until it reaches about 100
million degrees Kelvin, when helium fuses through the 3α reaction. This process
spreads rapidly in the core since a degenerate gas is a good conductor, and the
helium burning rate increases rapidly. This rapid process is called the "helium
flash," and quickly drives the core to a high enough temperature that it becomes
non-degenerate, at which point it expands and cools. The star continues to burn
hydrogen in a shell and helium in its core.

Eventually, the helium in the core has all been converted to carbon, and the
3α process continues in a shell. The core again becomes degenerate, and the Sun
once again turns into a red giant. The Sun is unstable now since the helium burning rate is very temperature sensitive, and thermal pulses occur at a rate of every
few thousand years. The star develops a superwind associated with the pulses and
blows off its envelope, leaving a hot core behind. The envelope forms a planetary
nebula, an expanding ring around the remnant core. Given the mass of the Sun,
the pressure of gravitation will never cause the core of the Sun to get hot enough to
trigger carbon burning, so the Sun will become a white dwarf composed mostly of
carbon. The Sun, supported by a degenerate electron gas, will cool to a black dwarf.

6.3.2 Hertzsprung-Russell Diagrams

A Hertzsprung-Russell (HR) diagram is a plot of the luminosity of a star versus its
surface temperature. Such a plot for a collection of stars shows clearly that they
fall into classes. The main sequence is a class of stars reaching diagonally across
the HR diagram, representing stars burning hydrogen in their cores with a range
of masses. Low mass stars have a low luminosity and surface temperature, while
high mass stars are found in the high luminosity and high temperature region. As

stars evolve from protostars onto the main sequence and then age off the main sequence, a path is traced on the HR diagram as they change their characteristics. A vertical path on the HR diagram represents a radius increase, with an increase in luminosity due to increased surface area, at constant temperature. A horizontal path on the HR diagram (constant luminosity with an increase in temperature), represents a decrease in radius.

Chushiro Hayashi[2] of Kyoto University showed that stars contracting toward the main sequence have high luminosities due to high surface temperatures at large radii. The star has a large opacity and is highly convective in order to get the energy out fast enough. Hayashi showed that the star in this phase of contraction moves rapidly down a vertical path in the HR diagram until it is in the vicinity of the main sequence. These evolution paths are often referred to as Hayashi tracks or Hayashi lines. A Hayashi line drawn on an HR diagram represents the boundary between an allowed region on the left and a forbidden region on the right for stars of a given mass and composition that are in hydrostatic equilibrium.

As the Sun evolved from a protostar toward the main sequence, as seen in Figure 6.1, it initially moved vertically up the HR diagram as the core radius increased. As the temperature rises, the opacity decreases and the surface of the star moves in toward a smaller radius with an increasing temperature and constant luminosity. As the accretion of material slows, the luminosity declines, until the radius moves through the shock front, causing a sudden increase in the surface temperature, seen as the turn to the left in the HR diagram. The star's path now

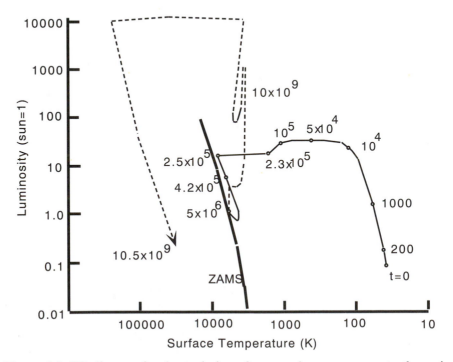

Figure 6.1: HR diagram for the evolution of a one solar mass star onto the main sequence and off again (after Winkler[4]). The approximate times in years are given for the evolutionary process steps.

turns down in the HR diagram, with luminosity and effective radius declining until the onset of hydrogen fusion. The star then settles onto the main sequence for the majority of its life. At the end of the Sun's long life on the main sequence, changing to shell burning, it will move toward a larger radius and luminosity as it becomes a red giant. When the helium burning transition is made, there is a rapid drop in luminosity and radius of the core, followed by a red giant phase again as hydrogen and helium shell burning occurs. As the star blows off its envelope at high luminosity, the star's radius decreases and its surface temperature increases dramatically, seen as the sharp turn to the left in the HR diagram near the end of the life of the star. This is followed by cooling and the formation of a white dwarf.

6.4 Computational Approach

6.4.1 Evolution of the Present Model Starting at t = 0

The star model begins from a very diffuse ($\rho = 10^{-19}$ g/cm^3) gas cloud of uniform density and composition. The cloud collapses due to gravity and is governed by the laws of ideal gas thermodynamics. The free-fall time for a cloud is approximately $\tau \approx (3G\rho)^{-1/2}$, which for $\rho = 10^{-19}$ g/cm^3 is about $\tau \approx 400,000$ years. This time is long compared to the thermal equilibration time, so the collapse occurs isothermally. A significant simplification is made in this model by assuming that the cloud is made of atomic gas rather than dust and molecules, since the evaporation of the dust and the excitation of molecules plays a role in the energy absorption of the collapsing cloud. The gravitational collapse of the cloud is simplified as a first-order computation.

As the cloud collapses, gravitational potential energy is converted into thermal energy, and the interior temperature rises. The model protostar, starting from the diffuse cloud, is treated as a series of equal mass layers which collapse uniformly (with no rotation) and do not mix. The collapse can be computed as free fall, with or without pressure effects and with or without temperature calculations (otherwise no change in temperature). When the interior temperature reaches a value where fusion begins, the star has reached the zero-age main sequence. The model cloud is simplified to be a mono-atomic gas, undergoing slow collapse.

6.4.2 Evolution of the Present Model Starting at the ZAMS

Starting from the main sequence star, the calculational approach is the same as that used for the static star program. A "shooting method" is used to integrate the differential equations (for T, ρ, r and L) from the center of the star outward and from the surface of the star inward to a "fitting mass point." The fitting point is $M/3$ by default (and can be changed in **ModifyModelParameters** under **Main-Sequence**). This integration is repeated with a variation in each of the four boundary parameters ($T_{central}$, $\rho_{central}$, $T_{effective}$, and L_R). The deviation in the four values (T, ρ, r, L) at the fitting point, with respect to variation in the boundary parameters, gives corrected values of the four boundary parameters. The iteration is repeated until the model converges, giving hydrostatic and thermal equilibrium.

The composition (X, Y, Z) of the star is stored into N radial layers (50 layers are used in the code). When the model has determined the stellar equilibrium condition for each time step, the composition is modified according to the burning rate in each layer. The maximum burn rate in all the layers of the star is found, and a fraction of the time required to consume the shell with this maximum rate is determined. This burn fraction per time step is 1% by default, and can be modified under **Modify Model Parameters** under **Main Sequence**. This gives an evolution time step which is large for young stars and which gets progressively shorter as the star moves into shell burning. After the composition throughout the star is modified by this evolution step, the hydrostatic equilibrium is again found, and the process repeats until the model fails (which presently happens as the star moves into shell burning). The present simulation is a simplification in that it ignores the time-dependent components of Eq. 6.3.

6.4.3 Program Procedures

The protostar evolution is treated either as free fall or as free fall with thermodynamics in the Procedure **ProtoStarEvolution**. This is handled by computing the acceleration of each of the layers of the gas cloud and updating the position to first order and the thermodynamics of each layer. The protostar thermodynamics are very simplified using the virial theorem, with half of the change in gravitational potential energy for the layer going into internal energy. An ideal gas model is used to compute the temperature and pressure change. The program has the effects of pressure, temperature, and radiation turned off initially. If the user chooses to turn on pressure, the effect of pressure on the acceleration of each layer is included; otherwise, free-fall motion is assumed. If the user chooses to turn on temperature effects, the temperature of each layer is changed by the virial theorem effect; otherwise, the temperature is a constant 2K. If the user chooses to turn on radiation, the luminosity and temperature are decreased by the loss to radiation from each layer. The energy lost to radiation is accounted for in a simplified manner, with no absorption in other layers. The protostar evolution is halted when the initial gas cloud has collapsed to a radius of one solar radius, or the central temperature exceeds 100 million degrees Kelvin. At this point, the star's central temperature has become high enough that hydrogen begins to fuse, and the star becomes a ZAMS star. This protostar model is extremely simplified, giving the user some feel for collapse effects, but could be made much more sophisticated by a user who is willing to modify the code.

The evolution starting from the ZAMS is controlled by the Procedure **Compute**. **Compute** initializes the model by calling **PrepareForModelCalc,** and then repeatedly calls **StellarModel,** which uses the same relaxation process described in the Interior Model of a Star computational approach, to relax the present stellar conditions. After **StellarModel** has gone through a successful relaxation, it calls **ModifyComposition,** which changes the hydrogen, helium, and heavy element mass fraction in each of the star's layers based on the fusion process rate occurring there. The star ages at a rate determined by the fastest fusion rate, and is limited to keep the fractional change in composition to less than a 1% change. The evolution would continue, but the present model fails to find a valid internal solution during hydrogen shell burning. The computer model needs to be improved to move beyond this limit.

6.5 Exercises

The following exercises are a few of those possible with the model. Some extensions to the code are described later to expand the model for other investigations of stellar evolution. Use of the program should begin with exploration of the menu selections to become familiar with the program capabilities (see Running the Program below).

6.1 The Sun as a Protostar

The Sun is the most obvious initial model to investigate, and the program defaults for the initial boundary conditions are those for a Sun-like star. Run the **Protostar Evolution** model from the **Protostar** menu and watch the evolution. Note the time for the stellar collapse. Is the collapse homologous? The cloud is seen to collapse from the inside out. Using the **Modify Model Parameters** in the **Protostar** menu, turn on the pressure, temperature, and radiation effects one at a time, and rerun the free fall. When pressure is turned on by itself, the inner part of the cloud collapses, but the outer part is prevented from falling inward by pressure. You will need to stop the model, since it will never reach a stopping condition. What is the effect of turning on the temperature and radiation effects? What is the effect on the free-fall time? Compare the HR diagram result with that given by Figure 6.1. The crudeness of the very simplified free-fall model used here shows in this comparison.

6.2 The Sun Starting From the ZAMS

The program defaults for the initial boundary conditions are those for a zero-age Sun-like star. Run the model from the **Main Sequence** menu. You may want to stop after about 16 billion years, since the model has trouble moving beyond this age. Compare the result for the HR diagram with that given by Figure 6.1. What has happened to the pp and CNO production regions of the Sun? What is the helium and hydrogen distribution? Try to help the program move the Sun beyond 16 billion years by modifying the model parameters and using the hot keys.

6.3 High Mass Stars

Explore the protostar evolution of a high mass star (e.g., M = $50 M_{sun}$). Since the initial density in the model is a constant independent of star mass, does the protostar collapse time differ with the star's mass? Look at the evolution starting from the ZAMS. How good is the agreement for the HR diagram with that given by Clayton[1] (chapter 6). What is missing from this model that contributes to the different result?

6.4 Low Mass Stars

Explore the evolution of a low mass star (e.g., M = $0.2 M_{sun}$) for both the protostar and ZAMS star. How good is the agreement with that given by Clayton[1] (chapter 6). What is missing from this model that contributes to the different result?

6.5 **Free Fall**

It was previously stated that the free-fall time was $\tau \approx (3G\rho)^{-1/2}$. Derive this relationship calculating the time for a mass to fall to $r = 0$ due to the gravitational attraction of the collapsing cloud. Hint: This can be done by placing all the star's mass at the origin, and a test mass at a the cloud's radius as determined by the density. Compare the result with the time coming from the simulation.

6.6 **Main Sequence Lifetime**

Use the model to find the main sequence lifetime for stars of various mass (e.g., masses of 0.5, 1, 2, 10, 100) and compare these with theory.

6.7 **Final Star Cooling Rate**

At the end of the sun's life, it will slowly cool from a white dwarf toward a black dwarf, as shown in the HR diagram generated by the **Stages M=1 Star Evolution** menu selection. Determine what the slope of this cooling line should be for the Sun.

6.8 **Star Data**

The data values shown in the **Show Star Data HR Diagram** menu selection are contained in the file EVOLVE.DAT. The data at the top of this table are for actual stars, and the rest are calculated. Locate ten more star data values and add them to this table. The data values for each star in this file are X, Z, mass, temperature, luminosity, central temperature, central density, radius, and a star color flag, which is 1 for actual stars and 0 for computed stars.

6.9 **Protostar Thermodynamics**

Make a plot of temperature, luminosity, and pressure versus radius for an adiabatically collapsing gas cloud.

6.10 **Hayashi Tracks**

For a mass 1 star, determine and plot on an HR diagram the motion of the Hayashi tracks for stellar evolution. Compare this with the model calculation for a mass 1 protostar.

6.11 **Chemical Composition**

How does the chemical composition affect the evolution of a star off the main sequence? Set the composition to very hydrogen rich, and compare the time to evolve. Repeat for a hydrogen-poor star.

6.6 *Possible Modifications to the Program*

There are a number of extensions that can be made to this model to improve and broaden the physics content. The following is a partial list of these possible extensions.

6.6.1 Plot Parameters Versus Radius

The STELLAR code includes an option to plot the various parameters versus radius instead of mass. Modify this code to include this option. Examine the STELLAR code to see how this is done by searching for the variable called **PlotMassOrRadius.**

6.6.2 Main Sequence Evolution

The present program reaches a point evolving from the ZAMS where it gets stuck in its evolution, usually as the star moves into hydrogen shell burning. Work out a technique in the software (Procedure **Compute**) to break the program through this barrier, which results from the inability to match all the boundary conditions at the fitting point.

6.6.3 Better Protostar Model

The present protostar collapse model is very simplistic, with first-order integration of the motion and insufficiently detailed thermodynamics. Modify the **ProtostarEvolution** routine to improve the protostar model with better thermodynamics and opacity effects on luminosity. The difficulty in creating a sophisticated model includes instability of the dynamics, and the strong non-linearity of effects such as radiation.

6.6.4 Heavy Element Burning

Add variables to account for the burning of elements heavier than helium in the star (carbon, oxygen, ...), and add these burning processes to the energy generation routine **NuclearPower.**

6.6.5 Protostar Composition

Add to the program the handling of dust and molecular gas in the handling of energy usage and opacity in the collapsing star.

6.6.6 Added Time Dependence

Add in the time dependence in the calculation from Eq. 6.3 for luminosity, which is presently left out.

6.7 Details of the Program

6.7.1 Running the Program

This section describes the menu options found in the program. The main menu bar is arranged so that the **Boundary**, **Protostar**, and **Main Seq** menus contain the main model computation, while the other main menu items provide related information. A menu item **Getting Started** is provided under the **File** menu to help guide your initial usage of the code.

- **File**: Disk file handling.

  - **About EVOLVE**: Help screen about this program.

  - **About CUPS**: Help screen about the CUPS project.

  - **Getting Started**: This provides a help screen giving instructions for getting started with the program.

  - **File Save of Results**: Writes a disk file of the calculation results containing printable text. The file name begins with the letters "ES" and contains the star's mass value. The four fitted parameters are included as well as the values of mass, temperature, density, radius, luminosity, opacity, power, and pressure stepping out through the star. This feature is provided so that model data can be used outside this program, or re-read by this code.

  - **File Read of Previous Results**: Reads a disk file of the calculation results previously written, allowing the continuation of a calculation.

  - **Configuration**: Allows for modification of the screen colors, changing the temp directory, and providing memory information.

  - **Exit**: Exit the program.

- **Stages**: This menu gives selections related to the stages of evolution.

  - **M=1 Star Evolution**: Shows schematically in time the steps in the evolution of a one solar mass star. The code shows a series of evolution steps, moving to the next step when the mouse is clicked, or any key is entered. The **Esc** key can be used to exit the sequence. Figure 6.2 shows the screen at the end of this sequence.

  - **Explain Evolution**: This selection gives a brief descriptive screen of the stages of evolution.

- **HR Diagram**: This menu provides items related to Hertzsprung-Russell diagrams.

  - **Show Star Data HR Diagram**: An HR diagram is plotted for a set of star data of a range of masses (contained in disk file EVOLVE.DAT). Some data is actual star data, while most of it is the result of calculations. Figure 6.3 shows the screen resulting from this selection. Note the main sequence, red giant, and white dwarf regions. One of the exercises is to increase this data set with more actual star observations.

  - **Show HR Diagram Trends**: An HR diagram which shows schematically the trends for stars in the diagram. Any keystroke or mouse click steps through the demo sequence. It shows the ZAMS line, the Hayashi limit lines, and lines indicating the dependence of luminosity on radius.

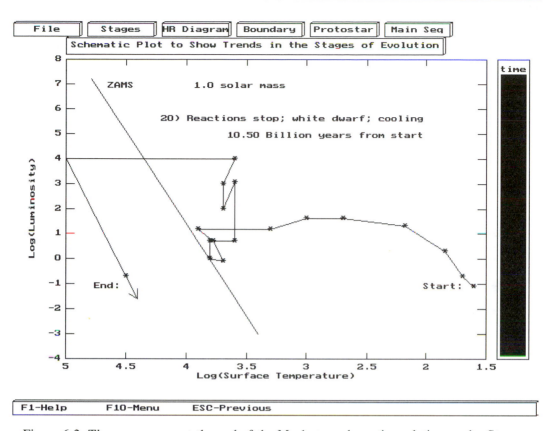

Figure 6.2: The screen seen at the end of the M=1 star schematic evolution, under Stages.

- **Explain HR Diagram**: This selection gives a brief descriptive screen of the meaning of an HR diagram.

• **Boundary**: This selection provides for the input of the boundary condition parameters for the model.

- **Modify Initial Parameters**: Change the parameters for the model, just as in the Interior Model of a Star program. The mass is used as a fixed value, while the four values central temperature, central density, luminosity, and surface temperature are initial guesses for the values that will be determined by the model. These parameters are used by the **Main Sequence** computation. There is a table of initial guesses given at the end of chapter 5.

- **Modify Chemical Compositions**: This selection allows for setting the initial values of the composition X, Y, and Z in the model. The values of X and Y are entered here, and Z is determined from the constraint that the sum $X + Y + Z$ must equal 1.

- **Reinitialize Program**: This reinitializes the software to its initial default parameters when the program was started.

• **Protostar**: This menu provides selections related to protostar formation and evolution.

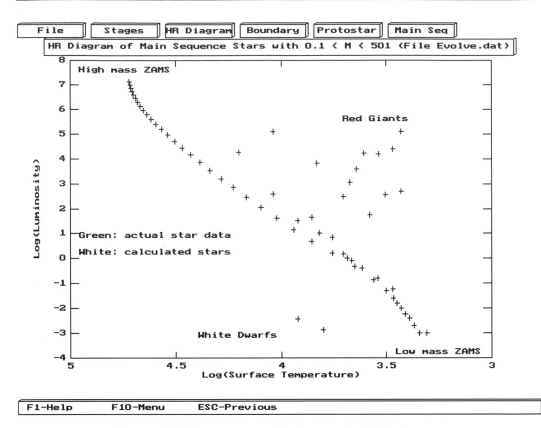

| File | Stages | HR Diagram | Boundary | Protostar | Main Seq |

HR Diagram of Main Sequence Stars with 0.1 < M < 50! (File Evolve.dat)

Figure 6.3: The screen seen for the star data H-R diagram.

– **Protostar Evolution**: Shows the evolution of a protostar cloud under gravitational free fall. Figure 6.4 shows the screen partway through the time evolution of the cloud. The menu item **Modify Model Parameters**, described below, allows for controlling options which affect the model calculation. Try turning on these effects one at a time to observe the impact on the protostar collapse.

– **Show Protostar Results**: Shows a table of the present calculation results for the star's parameters as a function of mass stepping out from the center of the star.

– **Modify Model Parameters**: Adjusts the protostar model parameters, including step time, and flags for pressure, temperature, and radiation effects. The step time is the size in years of each evolutionary step. The model defaults to a condition of free fall, where there is no effect from pressure, temperature, or radiation upon the collapse. By turning on these effects one at a time, you can gain some understanding of how the physics of the collapse is affected. The collapse will always start out from the inside to the outside. This is a very simplified model for protostar formation.

– **Explain Protostar**: This selection briefly explains the protostar formation and evolution.

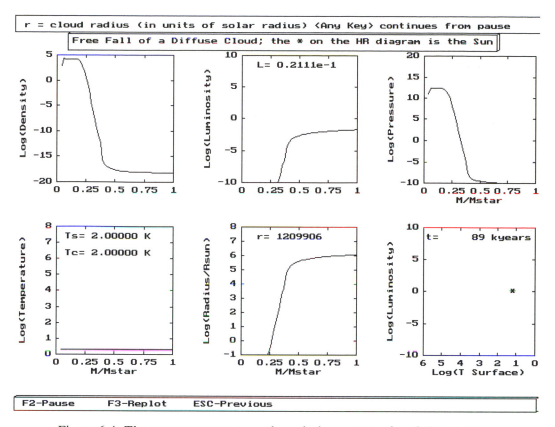

Figure 6.4: The screen seen partway through the protostar free-fall evolution.

- **Protostar Thermodynamics**: This selection shows the protostar thermo-dynamics result of the adiabatic assumption.

- **Main Sequence**: This selection is used to initiate the computation of the model from the ZAMS.

  - **Run One Model**: Starts the iterative calculation of the model for one time step. The program quits iterating if, after 1000 iterations, the model has not converged. The **Esc** key can be used to abort the iteration *after* the present loop is finished. The program shows four plots on the screen as it iterates: log(*Density*) versus (M_r/*Mass*), log(*Temperature*) versus (M_r/*Mass*), log(*Luminosity*) versus (M_r/*Mass*), and log(*Radius*) versus (M_r/*Mass*).

  - **Evolve From ZAMS**: Initiates the evolution of the star through time. It performs the same relaxation of **Run Model** above for an instant in time, followed by a composition change. The composition change is based upon the nuclear burning rate in each of shells of the star and upon the time step chosen for each evolution step. The model will iterate forever. At some point in time, the model fails to converge, usually during the shell burning phase. The program shows in a message box the last Delta values for each parameter, which are the fractional differences between the inward and outward integrations. These parameters must be within the error value set in the menu

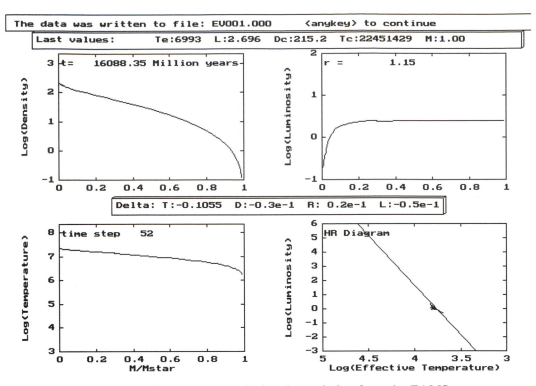

Figure 6.5: The screen seen during the evolution from the ZAMS.

item **Modify Model Parameters** in order to move to the next time step. A data file beginning with "EV" plus the star's mass value is written on each time step as a record of the evolution (e.g., EV001.000 for a mass one star). This file is recreated each time this option is selected. Figure 6.5 shows the screen part way through the time evolution from the ZAMS. There are several "hot keys" available during the convergence loop. Only one keypress is read per loop, so don't expect response to multiple keypresses. The program beeps whenever it accepts a hot key input. The hot keys are

* **'c'**: Decreases the present boundary value of the central temperature by 10%.

* **'C'**: Increases the present boundary value of the central temperature by 10%.

* **'d'**: Decreases the present boundary value of the central density by 10%.

* **'D'**: Increases the present boundary value of the central density by 10%.

* **'e'**: Decreases the present boundary value of the effective temperature by 10%.

* **'E'**: Increases the present boundary value of the effective temperature by 10%.

* **'l'**: Decreases the present boundary value of the luminosity by 10%.

* 'L': Increases the present boundary value of the luminosity by 10%.

* '+': Increase by a factor of 2 the percentage used as the delta in the boundary condition changes above (10% by default).

* '-': Decrease by a factor of 2 the percentage used as the delta in the boundary condition changes above.

* 'p' or 'P': Pause the computation (restart with any key).

* 'r or 'R': Redraws the screen plots.

* 's or 'S': Stop the evolution after the next convergence.

– **Continue Evolution**: Continues an evolution started with the **Run Evolution** option, appending the results to the disk file.

– **Show Model Results**: Shows a table of some of the numerical results from the calculation for mass, temperature, density, radius, and luminosity stepping out in mass from the center of the star.

– **Show More Results**: Shows a table of some more of the numerical results from the calculation.

– **Plot Model Results**: Plots the results of the last computation, as done during **Run Model**.

– **Plot More Results**: Plots more of the results of the last computation, including opacity, power and pressure versus mass, as well as an HR diagram.

– **Plot Composition**: Shows the distribution of hydrogen (X), helium (Y), heavy elements (Z), and nitrogen in the star at the present point in the evolution. Note the depletion of hydrogen at the center of the star.

– **Modify Model Parameters**: Adjusts the fitting criteria. The value of the precision of the fit to the star mass at the "fitting mass point" may be adjusted, as can the model precision (fractional delta in the match of the four integrated parameters at the fit point) and the location of the "fitting mass point," and the fractional composition change at each time step. The sound can be enabled and disabled, producing a beep at the end of each convergence cycle.

6.7.2 Structure of the Program

The program Evolve consists of one large program, EVOLVE.PAS, and one unit, EVOLVEC.PAS. The following gives the program structure in terms of the menus:

Main Menu

  – File

    * About EVOLVE—About This Program

    * About CUPS—About the CUPS Project

* Getting Started—Guide to Initial Usage of the Program
* File Save—Save Results to File
* File Read—Read Results From File
* Configuration—Change or Examine the Program Environment
* Exit—Exit Program

– Stages—Show Stages of Evolution

* M=1 Star Evolution—How a Mass 1 Star Evolves
* Explain Evolution—Describe Evolution Process

– HR Diagram—Describe HR diagrams

* Star Data HR Diagram—Show HR Diagram for Set of Stars
* HR Diagram Trends—How Stars Evolve on an HR Diagram
* Explain HR Diagram—Describe an HR Diagram

– Boundary—Enter the Mass and Boundary Initial Conditions

* Modify Parameters—Modify the Model Parameters
* Composition—Control the Stellar Composition
* Explain Conditions—Describe the Stellar Conditions
* Reinitialize—Reinitialize to Program Defaults

– Protostar—Perform Protostar Calculations

* Protostar Evolution—Show Protostar Free Fall
* Show Results—Display Model Results
* Modify Parameters—Change Protostar Model Parameters
* Explain Protostar—Describe Protostar Formation and Evolution
* Protostar Thermodynamics—Explore the Thermodynamics of a Protostar

– Main Sequence—Perform Iterative Computation and Show Graphical Results

* Run One Model—Run the Model Calculation
* Run Evolution—Run the Evolution of the Model Calculation
* Continue Evolution—Continue the Evolution of the Model Calculation
* Show Results—Display Model Results
* Show More Results—Display Model Results
* Plot Results—Plot Model Calculation Results

* Plot More Results—Plot Model Calculation Results

* Plot Composition—Plot Stellar Model Composition

* Modify Parameters—Modify Model Parameters

Acknowledgments

Some of the approach to the computation used here is based on that developed by Bohdan Paczynski of Princeton University and is greatfully acknowledged.

Bibliography

1. Clayton, D. *Principles of Stellar Structure and Nucleosynthesis.* Chicago: University of Chicago, 1983.

2. Hayashi, C. Publication of the Astronomical Society of Japan **13:**450, 1961.

3. Kippenhahn, R., Weigert, A. *Stellar Structure and Evolution.* Berlin: Springer-Verlag, 1990.

4. Winkler, K. A., Newman, M. J. *Astrophysical Journal* **26:**201, 1980.

7

Stellar Pulsation

Charles A. Whitney

7.1 Introduction

7.1.1 The Importance of Stellar Pulsation

The surfaces of most, and perhaps all, stars oscillate in a gentle pattern of irregular waves. These waves have little more effect than ripples on an ocean, but some stars also undergo a much more spectacular type of motion that may have a profound influence on the life of the star. The layers of pulsating stars swing in and out in a global pattern that repeats in periods from a few seconds to a few years, depending on the time taken for a sound wave to travel from the center to the surface of the star. (Due to the effects of gravity, the pulsation period is several times longer than this interval).

Pulsating variables are divided into about a dozen classes, depending on their location on the HR diagram and their physical location in galaxies. (See ref. 7 for an introduction to this topic.)

The details of the pulsation can, in the context of a theory, tell us something about the interior of a star. Aside from the neutrino radiation from the Sun, these patterns of pulsation are one of the few reliable techniques for probing stellar interiors. Hence, they occupy a central position in theoretical and observational astrophysics. Pulsating stars are also used for determining distances in the universe, so they are important to cosmologists, as well.

What the Program Will Do

This simulation is intended to be used by instructors to generate demonstrations for introductory astronomy courses and by students in upper-level undergraduate astrophysics courses. It illustrates stellar pulsation by simulating the thermomechanical behavior of a "star" modeled by a self-gravitating gas divided by spherical elastic shells. The elastic shells resemble a set of coupled oscillators. The program solves for the modes of small-amplitude motion, and it uses Fourier synthesis to construct motions for arbitrary starting conditions. The screen displays the thermodynamic structure and surface properties, such as temperature, total radiation output, and velocity.

Animation displays the nature of the pulsation; by showing the motions, temperatures, and energy flux, the program demonstrates the heat engine acting inside the pulsating star. The motions of the shells and the spatial Fourier decomposition into eigenmodes are displayed simultaneously, and this will help the students visualize the meaning of the Fourier components.

The program can illustrate the main cause of pulsation in terms of the behavior of heat engines. With suitable adjustments of the stellar parameters, the program can also be used to survey the differences between various groups of pulsating stars.

The discussion in this chapter will concentrate on classical Cepheids, although the RR Lyrae stars, W Virginis stars, and Mira variables can probably be studied with the aid of this program. White dwarfs are outside the scope of this program, as their equation of state and heat transfer mechanisms are not presently included in our equations.

7.2 The Conservation Equations

The equations describing the gas in chapter 4 on static stellar interiors must be elaborated to include time-dependent terms. The equations describing the radiation, on the other hand, do not need modification, and we can adopt the radiative diffusion approximation.

7.2.1 Conditions Assumed During the Pulsation

In this simulation, the model is assumed to remain spherical throughout the pulsation. The stellar layers oscillate in and out with a velocity that depends only on distance from the center of the star and is much smaller than the speed of sound. (This assumption is valid in most of the star, and it permits us to study the cause of the pulsation, but a reliable study of the atmosphere and the emission of light from the star must include shock waves.)

The gas motions resemble the free-swinging oscillations of air in an organ pipe or, in a cruder analogy, the swinging of a chain suspended from a hook. The open end of the pipe, or the free end of the chain, represents the surface of the star, and the hook, where the chain is held motionless, represents the geometric center of the star. (Spherical symmetry dictates that the displacements and velocities vanish at the center.) We shall now write the equations describing the interior motions of the model. (For historical discussions of this topic see refs. 4, 7, 18, 25, 26).

7.2.2 Mass

In place of a fixed spatial coordinate (Eulerian system), we use interior mass, $M(r)$, as the independent spatial variable (Lagrangian system). We attach a coordinate system to the matter, and let $r(M, t)$ be the radial distance of a mass element at time t, given by

$$M(r) = \int_0^r 4\pi\rho r^2 \, dr. \tag{7.1}$$

The differential of distance is given by

$$dr = \frac{dM}{4\pi\rho r^2}, \tag{7.2}$$

and the specific volume is defined by

$$V \equiv \frac{1}{\rho} = 4\pi r^2 \frac{dr}{dM}. \tag{7.3}$$

7.2.3 Momentum

Newton's law is

$$\frac{\partial^2 r}{\partial t^2} = -\frac{GM(r)}{r^2} - 4\pi r^2 \frac{\partial P}{\partial M}. \tag{7.4}$$

The first term on the right is the gravitational acceleration, and the second term is the acceleration produced by the pressure gradient.

7.2.4 Energy

The second law of thermodynamics is $c_V \, dT = T \, dS$, where c_V is the specific heat at constant volume, T is the temperature, and S is the entropy. Since the luminosity $L(M, t)$ is the flow of energy across a sphere of radius $r(M, t)$, this may be expressed as

$$c_V \frac{dT}{dt} = T \frac{dS}{dt} = -\frac{\partial L}{\partial M}. \tag{7.5}$$

The Lagrangian derivative with respect to time follows the matter, while the partial on the right is taken at an instant of time.

We evaluate the luminosity from

$$L = (-4\pi r^2)^2 \frac{4\sigma_B}{3\kappa} \frac{\partial T^4}{\partial M}. \tag{7.6}$$

We have written σ_B for the Stefan–Boltzmann constant and κ for the Rosseland mean opacity.

7.2.5 Equation of State and Opacity

The gas pressure is given by

$$P = \frac{\mathcal{R}\rho T}{\mu}, \tag{7.7}$$

where $\mathcal{R}$ is the gas constant, T is the temperature, μ is the mean molecular weight, and ρ is the gas density. The mean molecular weight is defined in terms of the total number density of particles, N, by

$$\mu \equiv \frac{\rho}{N}. \tag{7.8}$$

Calculation of N requires evaluating the fractional ionizations of all the species in the gas. Because the ionization energy is so much greater than the average thermal energy, we can assume without loss of accuracy that each species can only exist in two stages of ionization at each point. We define the fractional ionizations for hydrogen and helium:

$$xH = \frac{N_{H^+}}{N_{H^+} + N_H}, \qquad xHe = \frac{N_{H_e^+}}{N_{H_e^+} + N_{H_e}}, \qquad yHe = \frac{N_{H_e^{++}}}{N_{H_e^{++}} + N_{H_e^+}}. \tag{7.9}$$

The fractional ionizations are evaluated from the Saha-Boltzmann equation, and are carried out by **Procedure DoState1**. See, for example, refs. 6, 25. We simplify the calculation by considering the temperature in three ranges:

- Low: find xH with $xHe = 0$, $yHe = 0$;

- Medium: $xH = 1$, find xHe with $yHe = 0$;

- High: $xH = 1$, $xHe = 1$, find yHe.

Two other procedures call on DoState1:

- **Procedure FindV** takes P, T, and chemical composition and carries out an iterative solution to find V and the fractional ionizations.

- **Procedure FindT** takes P, V, and chemical composition and carries out an iterative solution to find T and the fractional ionizations.

7.2.6 Thermodynamic Differentials

If $\mathcal{Q}(\rho, S)$ is any of the thermodynamic variables considered a function of density and entropy, we may write its differential as

$$d\mathcal{Q} = \left(\frac{\partial \mathcal{Q}}{\partial \log S}\right)_\rho d \log T + \left(\frac{\partial \mathcal{Q}}{\partial \log \rho}\right)_s d \log \rho. \tag{7.10}$$

The adiabatic exponents are defined by

$$\Gamma_1 \equiv \left(\frac{\partial \log P}{\partial \log \rho}\right)_s, \qquad \frac{\Gamma_2 - 1}{\Gamma_2} \equiv \left(\frac{\partial \log T}{\partial \log P}\right)_s, \qquad \Gamma_3 - 1 \equiv \left(\frac{\partial \log T}{\partial \log \rho}\right)_s. \tag{7.11}$$

The program uses expressions for the adiabatic exponents in an ionizing gas found in ref. 6.

7.3   Static Model

We must first build a spherical, static model of a stellar interior by setting the time derivatives equal to zero in the conservation equations. The pulsation of the model is not sensitive to conditions in the deep interior, so we can assume a small rigid core at the center. This means we do not need to integrate the static model from the center outward and the surface inward; we merely integrate inward from the surface and cut off the integration when a considerable fraction of the star's mass has been included. Thus the model can be built without iteration. We assume that the model is chemically homogeneous.

The momentum equation becomes the equation of hydrostatic equilibrium,

$$dP = \frac{GM(r)\,dM}{4\pi r^4}. \tag{7.12}$$

We also need equations to define the density or temperature structure.

7.3.1   Constant-Density Model

The simplest approach to the static model is to assume $\rho = constant$. This constant-density model is highly unrealistic, but very useful. The hydrostatic equation may be integrated in closed form and the pulsations may be treated analytically (see, for example, ref. 18). The behavior of this model makes a useful demonstration of the influence of the high central concentration found in more realistic models.

All shells will have the same density, so we may integrate Eq. 7.1 to find

$$M(r) = \frac{4}{3}\pi r^3 \rho, \tag{7.13}$$

where

$$\rho = \frac{3M(R)}{4\pi R^3}, \tag{7.14}$$

and the pressure follows from integration of Eq. 7.12 pulse, noting that the pressure vanishes at the surface.

Having found density and pressure, we may find all other thermodynamic variables from the equation of state, Eq. 7.7. In passing, we note that the hydrostatic equation for any model for which the density is given as $\rho(r)$ may be integrated to find the pressure, temperature, etc.

7.3.2   Radiative and Convective Models

An alternative to arbitrarily prescribing the density is to specify the temperature gradient. If the model is radiative we have from Eq. 7.6

$$L = -(4\pi r^2)^2 \frac{4\sigma_B}{3\kappa}\frac{dT^4}{dM} = constant. \tag{7.15}$$

On the other hand, if the region is convective and has an isentropic structure, we have from Eq. 7.11

$$\frac{d \log(T)}{d \log(P)} = \frac{\Gamma_2 - 1}{\Gamma_2}, \tag{7.16}$$

so

$$\frac{d \log(T)}{dM} = \frac{\Gamma_2 - 1}{\Gamma_2} \frac{1}{P} \frac{GM(r)}{4\pi r^4}. \tag{7.17}$$

The program also permits construction of a model in which the radiative and convective gradients of temperature are compared at each depth and the lesser of the two is adopted. This is the conventional approach to stellar interiors where convection is assumed to be very efficient due to the high density of the gas. In a stellar atmosphere, the gas density is too low to ensure such high efficiency, so the atmosphere program in chapter 7 uses a mixing–length theory to evaluate the efficiency.

7.4 *Computational Approach*

The next step is to write the equations in a computational form, and we will follow the method of Castor,[4] with a slight change of the indexing.

7.4.1 Finite Difference Formulation

Dynamical Equations

The stellar mass is layered in N homogeneous spherical shells numbered from 1 outward. The outer surface of the ith shell is called the ith interface. The radius of the interface is r_i. The mass interior to r_i is M_i. Variables defined at the interface are given integer indices; variables defined within the shell are given fractional indices.

We define two types of shell masses:

- $DM1_i = M_i - M_{i-1}$, the mass between r_i and r_{i-1},

- $DM2_i = \frac{1}{2}(DM1_i + DM1_{i+1}) = \frac{1}{2}(M_{i+1} - M_{i-1})$, the dynamical mass associated with the ith interface.

The innermost interface (index 0) is the stationary boundary at the bottom of the envelope. The masses of the shells follow a geometric progression in the program, except for the homogeneous model, where they are equal.

Newton's law for each shell is

$$\frac{d^2 r_i}{dt^2} = -\frac{GM_i}{r_i^2} - \frac{4\pi r_i^2}{DM2_i}(P_{i+\frac{1}{2}} - P_{i-\frac{1}{2}}). \tag{7.18}$$

The specific volume is

$$V_{i+\frac{1}{2}} = \frac{4\pi}{3} \frac{r_{i+1}^3 - r_i^3}{DM1_{i+\frac{1}{2}}}, \tag{7.19}$$

and the conservation of energy becomes

$$T_{i+\frac{1}{2}}\frac{dS_{i+\frac{1}{2}}}{dt} = \frac{L_i - L_{i+1}}{DM1_{i+1}}. \tag{7.20}$$

Static Models

Setting the time derivatives equal to 0 leads to the following equations for static models:

$$P_{i-\frac{1}{2}} = P_{i+\frac{1}{2}} + \frac{DM2_i}{4\pi r_i^4}GM_i, \tag{7.21}$$

$$r_i^3 = r_{i+1}^3 - \frac{3DM1_{i+\frac{1}{2}}}{4\pi}V_{i+\frac{1}{2}}. \tag{7.22}$$

For the homogeneous model, the values of $V_{i+\frac{1}{2}}$ are known, so these equations permit us to step from the surface inward.

For the radiative and convective models, the situation is more complicated because we do not know the value of density at the next step. It is connected to the pressure through the temperature and degree of ionization. Thus we must proceed by iteration, taking into account the energy transfer betwen layers. Because the stellar layers are so thick in this simulation, we must use care in writing the finite-difference expressions. For the radiative model, we have $L_i = constant$. We express L_i in terms of the temperatures of the adjacent layers with Stellingwerf's[24] expression:

$$L_i = (4\pi r_i^2)^2 \frac{4\sigma_B}{3DM2_i} < \frac{\delta W}{\kappa} >_i, \tag{7.23}$$

where

$$W_i \equiv T_i^4, \tag{7.24}$$

$$< \frac{\delta W}{\kappa} >_i = \left[\left(\frac{W}{\kappa}\right)_{i+1} - \left(\frac{W}{\kappa}\right)_i \right] \Big/ \left[1 - \frac{\log(\frac{\kappa_{i+1}}{\kappa_i})}{\log(\frac{W_{i+1}}{W_i})} \right]. \tag{7.25}$$

This expression contains the temperatures of adjacent layers and it also contains the opacity κ, which is a function of density and temperature. There is no way to solve directly for temperature, so we proceed by guessing a temperature, evaluating the resulting error in the luminosity and then adjusting the guessed temperature until the luminosity comes out right.

Finally, for the convective model, we have the isentropic gradient Eq. 7.11. At each step we work by iteration. We guess the next T_i from the known pressure and an approximate value of Γ_2; we then find the resulting density and a new value of Γ_2; this leads to a new value of the temperature, and we repeat this process to convergence.

7.4.2 Linearization of the Dynamical Equations

We now develop a scheme for handling the time-dependent equations. The key step is "linearization," which leads to a set of equations in which each time-dependent variable appears only to the first power. This makes possible an analytical solution and prepares the way for the modal analysis and Fourier synthesis. The linearized equations give an excellent description of the pulsation in the envelope, although it fails in the atmosphere. Thus, it provides a good estimate of the periods and the growth or decay of the pulsation amplitude, but it is not reliable for the observed light or radial velocity curves.

The linearization process consists of writing each variable as the sum of a static value and a time-dependent term.

For example, the position of an element is

$$r(M_i, t) = r(M_i) + \delta r(M_i, t). \tag{7.26}$$

Similar expressions are written for the other variables, and they are inserted into the equations of motion. The time-variations are assumed to be small, so terms containing products of the variations, such as $\delta P \delta r$, can be ignored, and we replace terms like $\delta \log \rho$ by $\delta \rho / \rho$. This leads to

$$\frac{\delta T}{T} = \frac{dS}{c_V} + (\Gamma_3 - 1) \frac{\delta \rho}{\rho}, \tag{7.27}$$

$$\frac{\delta P}{P} = \frac{\rho(\Gamma_3 - 1)}{P} T \, dS + \Gamma_1 \frac{\delta \rho}{\rho}. \tag{7.28}$$

In order to simplify the solution of the resulting equations, we replace δr_i by

$$X_i \equiv (DM2_i)^{\frac{1}{2}} \delta r_i, \tag{7.29}$$

so

$$\left(\frac{\delta \rho}{\rho}\right)_i = \frac{\rho_i}{DM1_i} \left[\frac{4\pi r_i^2}{(DM2_i)^{\frac{1}{2}}} X_i - \frac{4\pi r_{i+1}^2}{(DM2_{i+1})^{\frac{1}{2}}} X_{i+1} \right]. \tag{7.30}$$

In place of the entropy variation we write

$$Y_i \equiv T_i \, \delta S_i. \tag{7.31}$$

X_i and Y_i become the new dependent variables, and all other variations may be found from them. Eq. 7.18 and Eq. 7.20 become (see ref. 4 for the G and K coefficients)

$$\frac{\partial^2}{\partial t^2} X_i = G1_{i,1} X_{i-1} + G1_{i,2} X_i + G1_{i,3} X_{i+1} + G2_{i,1} Y_{i-1} + G2_{i,2} Y_i, \tag{7.32}$$

$$\frac{\partial}{\partial t} Y_i = K1_{i,1} X_{i-1} + K1_{i,2} X_i + K1_{i,3} X_{i+1} + K1_{i,4} X_{i+2} + K2_{i,1} Y_{i-1} + K2_{i,2} Y_i$$

$$+ K2_{i,3} Y_{i+1}. \tag{7.33}$$

7.4.3 Modal Analysis

A model consisting of N shells can oscillate in N spherical modes. In a mode all layers move in and out with exactly the same period, and the layers come to rest at the same time. If all layers reach their greatest size at the same instant, this is the fundamental mode. In higher modes, called the first, second, etc., "overtone," the motions are synchronized, but some layers are moving inward while others are moving outward. Each mode has a different frequency of motion. These are the resonant (or eigen-) frequencies of the model. The higher modes are quicker than the fundamental, because different layers are pushing against each other rather than moving in the same direction. (In a musical instrument, this higher frequency is heard as a higher pitch.)

As an analogy, think of a clock pendulum. It has a single well-defined period that depends on the length of the pendulum and the acceleration of gravity. If we attach another pendulum to the end of the first, interesting things begin to happen: a second mode appears. In the fundamental mode, both pendulums swing to the same side; in the overtone, one moves to the right while the other moves to the left. This mode has a higher frequency than the fundamental. If we just hit the pendulum casually to start it swinging, we will find its motion rather complicated, because it is moving in both modes at the same time. But if we are very careful to start from rest and push each one with just the right impulse, we can excite a single mode. In this case the motions of the two pendulums are synchronized and the pattern is quite simple.

Rather than hunt for the right starting conditions by trial and error, it is possible to find them by solving a set of simultaneous algebraic equations. There are exactly as many equations as their are shells, and there are as many modes as there are equations. The theory is discussed in books on dynamics under a heading like "simple harmonic motion of coupled oscillators," and it is treated in the chapter by R. Jones in the Mechanics Course of this series, so we need not describe it in detail here (see also Cox[7]). The program solves these equations for all possible modes. (See ref. 17.)

Adiabatic Eigenfrequencies

If we assume no perturbation of the heat exchange between shells, $\delta L = 0$ and $\delta S = 0$, or $Y = 0$ everywhere. This permits canceling many terms from the equations of motion Eqs. 7.32 and 7.33. To find the modes, we assume that the variations may be written as the product of space-dependent and sinusoidal time-dependent terms, $X_i e^{i\omega t}$. (This form ensures that the motions will be synchronized.) Inserting these expressions into Eq. 7.32 we find, after some rearranging, N algebraic equations of the form

$$\omega^2 X_i = G1_{i,1} X_{i-1} + G1_{i,2} X_i + G1_{i,3} X_{i+1}, \ (i = 1, \ldots N), \qquad (7.34)$$

where the coefficients $G1_{i,j}$ are determined by the structure of the static model. There are N such equations. They lead to N values of ω_m^2. The eigenfrequencies are real, corresponding to undamped motion. Inserting each value of ω_m into Eq. 7.34 permits solving for the N values of displacement for each mode, $X_{i,m}$. The displacements are real, corresponding to motions in phase with each other.

Once the displacements $X_{i,m}$ are known, the other thermodynamic variables are found from Eqs. 7.27, 7.28, and 7.30.

7.4.4   Fourier Synthesis of Pulsation

Adiabatic

In the mth mode, the motion of the surface (N^{th} shell) is

$$X_{N,m}(t) = A_m \cos(\omega_m t + \phi_m), \tag{7.35}$$

and for the interior shells

$$X_{i,m}(t) = R_{i,m} A_m \cos(\omega_m t + \phi_m), \tag{7.36}$$

where we have defined the relative amplitudes by

$$R_{i,m} \equiv \frac{X_{i,m}}{X_{N,m}}. \tag{7.37}$$

The amplitudes A_m and the phases ϕ_m are fixed by the way in which the motion is started. (In this program, the user specifies the amplitude and phase of the first four modes.) Any general motion may be represented by a Fourier sum like the following, with suitably chosen coefficients:

$$X_i(t) = \sum_{m=1}^{N} X_{i,m}(t). \tag{7.38}$$

In the adiabatic case (no energy exchanged between layers) the coefficients and frequencies are all real. This means that the changes of all thermodynamic quantities are in phase with each other and continue with a constant amplitude. On the other hand, when thermal energy is exchanged between layers, during the pulsation, the coefficients become complex. The imaginary part of the frequency describes the growth or damping of the amplitude of that mode. If the amplitude grows, the mode is said to be unstable against pulsation, and we would expect the corresponding star to exhibit pulsations. The imaginary parts of the other coefficients produce phase shifts between the thermodynamic variables.

It is important to realize that while this linear theory can predict instability, the prediction is not accurate unless the heat transfer is treated in a fully non-adiabatic way. Furthermore, the linear theory cannot predict how far the pulsations will grow; the pulsation is ultimately limited by nonlinear effects, such as shock fronts, which are beyond the scope of this simulation.

7.5   Details of the Program

7.5.1   Running the Program

The user specifies the key parameters: mass, luminosity, and radius, as well as the chemical composition. (Helium, hydrogen, and a representative metal are included.) Once these parameters are set, the program executes an inward integration to find the pressure and temperature distributions inside the star's envelope. There are four alternatives for this calculation. The static model may 1) have a density that constant or it may be built on 2) radiative equilibrium or 3) convective equilibrium or 4) a mixture of the two.

Open the **EnterParameters** window, and you will see the default values corresponding to the Sun's envelope. Note the units of M, L, and R may be toggled between c.g.s. (centimeter gram second) and solar units with the radio button. You may alter any of the parameters by selecting the corresponding text box by moving the cursor to the box and pressing the mouse button. Then enter the new values from the keyboard. When the data are satisfactory, click the **AcceptData** button. The new value of surface temperature and gravity are displayed and the parameters are stored and ready for use by the program.

StaticModel/Build will carry out the integration from the surface inward to build the hydrostatic envelope of your star. The method of density calculation will depend on the type of model you have selected with the radio buttons at the bottom of the window. In all cases, the program solves the equation of hydrostatic equilibrium to find the pressure from the density. In the radiative model, the temperature gradient is adjusted to give constant radiative flux; in the convective model, the temperature gradient is evaluated from the adiabatic gradient. Due to simplifications made in calculating the equation of state and the opacity, the models are best for intermediate temperature (4,000 K to 50,000 K). In more extreme cases, the model may fail to converge. The program will alert you if a set of parameters appears doomed to failure or inaccuracy.

In building the model, the program usually truncates the last shell, and this reduces the envelope mass. Be sure you reset the mass to the desired value each time you rebuild a model, or you will find successive models having smaller envelope masses.

StaticModel/List will open a window and list the structure of the model.

StaticModel/Graph will construct graphs of physical variables in the static model envelope. You may select different variables with the popup menu in each graph, and you may select a logarithmic or linear y-scale for the variables P, T, and ρ.

To save the model for use at a later time select the **File/Save As...** item in the menu. You will be prompted for a file name. (The current model name is the default but you may change it from the keyboard.) If a file already exists with that name, you will be asked whether you wish to overwrite it. The file will be saved as a text file and you may open it and print it with a word processor. You may also open it with the **File/Open Model...** menu item, which will prompt you for the name of the file.

Sample Input and Output

Successful operation will only be achieved in a limited range of input parameters. The program examines your input and rejects outlandish values. The following will guide you in selecting appropriate values:

	Lower	Upper
Masses	0.1	100 Msun
Radii	0.01	1000 Rsun
Luminosities	0.01	10,000 Lsun
Num. shells	1	20
Helium abundance	0	0.99
Metals abundance	0.001	0.04
Envel. Mass	0.01	1

The following parameters were used for a mixed radiative and convective model of a Cepheid:

- Solar units: mass = 2.5, radius = 45, luminosity = 2500;

- Helium abundance = 0.30; metals abundance = 0.02;

- Initial envelope mass = 1.00;

- Number of shells = 10.

They produced the following model, which has a fundamental period of 10.649 days.
    These are the variables listed in the output:

- R radius of the shell (cm)

- M mass interior to shell (g)

- $T(K)$ temperature of shell

- $P(d/cm^2)$ pressure (including radiation pressure)

- $V(cm^3/g)$ specific volume, or reciprocal of density

- L radiant luminosity across shell (erg/sec/cm^2)

- κ opacity ($\kappa \rho dx$ is the increment of optical depth, so κ has units cm^{2}g^{-1})

- $dM1$ shell mass for structure (g)

- $dM2$ shell mass for dynamics (g)

- μ mean molecular weight (dimensionless)

- γ ratio of specific heats, including effect of radiation

- H^+ fraction of hydrogen in form of protons

- He^+ fraction of helium that is singly ionized

- He^{++} fraction of helium that is doubly ionized

radius,	intmass,	T,		P,	V,	flux,
kappa,	dm1,	dm2,	mu,	gamma, H^+	He^+	He^{++}

3.132e+12	4.970e+33	6.008e+3		5.912e+21	6.437e+8	9.567e+36
5.728e−2	4.309e+27	2.154e+27	1.321	0.303 0.003	0.000	0.000
3.109e+12	4.970e+33	1.773e+4		3.517e+3	6.722e+8	9.608e+36
5.876e+0	1.642e+28	1.037e+28	0.670	0.211 1.000	0.780	0.000
3.016e+12	4.970e+33	3.313e+4		1.612e+4	3.180e+8	9.583e+36
2.330e+0	6.260e+28	3.951e+28	0.661	0.261 1.000	0.949	0.051
2.830e+12	4.970e+33	5.045e+4		7.803e+4	1.073e+8	9.655e+36
3.348e+0	2.386e+29	1.506e+29	0.633	0.279 1.000	0.044	0.956
2.549e+12	4.970e+33	7.611e+4		4.367e+5	2.842e+7	9.597e+36
2.266e+0	9.095e+29	5.741e+29	0.631	0.295 1.000	0.001	0.999
2.182e+12	4.969e+33	1.200e+5		2.981e+6	6.422e+6	9.687e+36
2.376e+0	3.467e+30	2.188e+30	0.631	0.300 1.000	0.000	1.00
1.719e+12	4.965e+33	2.027e+5		2.817e+7	1.114e+6	9.636e+36
1.860e+0	1.321e+31	8.341e+30	0.631	0.307 1.000	0.000	1.000
1.160e+12	4.937e+33	3.575e+5		3.813e+8	1.382e+5	9.678e+36
1.242e+0	3.562e+31	2.442e+31	0.631	0.323 1.000	0.000	1.000
7.285e+11	4.815e+33	6.802e+5		6.747e+9	1.441e+4	9.651e+36
9.531e−1	1.047e+32	7.015e+31	0.631	0.336 1.000	0.000	1.000
2.982e+11	4.215e+33	1.846e+6		4.902e+11	5.266e+2	1.083e+37
6.252e−1	1.995e+32	1.521e+32	0.631	0.348 1.000	0.000	1.000
1.131e+11	4.612e+33	1.265e+7		6.442e+10	0.000e+0	0.000e+0
0.000e+0	8.000e−1	0.000e+0	0.631	0.400 1.000	0.000	1.000

Figure 7.1 shows a typical pattern of output.

7.6   *Period-Luminosity Relations*

The most remarkable observed property of classical Cepheids is the period-luminosity (P-L) relation, which was discovered by Henrietta Leavitt in 1911. She was studying the variable stars in the Magellanic Clouds (small nearby galaxies visible from the Southern Hemisphere), and she noticed that the brighter stars had longer periods of variation. In fact, when she made a plot of the data, she found a tight correlation between stars' average brightnesses and the durations of their cycles. All the stars in her study could be assumed to lie at the same distance from the Earth, and this observed relationship implied that the true brightness was correlated with period. Astronomers later realized that if they could determine the true distances of this group of stars, they could then turn the relation around and determine the distances of other galaxies. From the periods of the starts observed in another galaxy, they could determine the true brightnesses, and by comparison with the apparent brightness, they could infer the distance of the galaxy. A great deal of effort was required to calibrate this relationship between true brightness and period, because the distance to the Magellanic Clouds was poorly known at the time. In fact, the work of refining this relationship continues today. This procedure soon became our best way of determining the distances to remote galaxies and, in

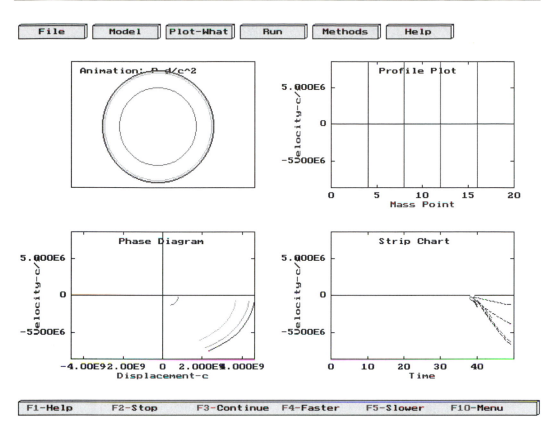

Figure 7.1: Illustration of four types of dynamic plots displaying the pulsation. Upper left: Animation showing the motions of the shells; upper right: profile plot showing the time-varying distribution of velocity in the interior of the star; lower left: phase diagram plotting the velocities in selected shells against the corresponding displacements; lower right: strip chart showing the velocities of selected shells. The user may select the type of plot and the variables to be plotted in each frame.

the 1920s, it lead to Edwin Hubble's announcement of the expansion of the universe. (Modern data for the Small Magellanic Cloud may be found in ref. 10. Also see ref. 14.)

W Virginis stars show a similar, but displaced, P-L relation. In contrast, the RR Lyrae stars all have nearly the same periods and the same luminosities (see ref. 15).

In the 1950s it became clear that the situation was more complicated than had been appreciated earlier; different groups of pulsating stars have slightly different P-L relationships. The RR Lyrae stars in different globular clusters, for example, have slightly different luminosities. And the Cepheids in the Magellanic Clouds show a scatter about the mean relationship that is probably not merely observational error. Such differences are associated with differences of chemical composition and age. (For theoretical discussions of pulsational instability and the period-luminosity relation see refs. 2, 5, 8, 9, 26).

7.7   The Hertzsprung Relation

Successful tests of the theories of pulsation and of stellar development depend on observing properties of stars that can be evaluated quite accurately from the Earth. For example, tests based on the period-luminosity relation are difficult to carry out because deriving true luminosities depends on knowing distance and being able to account for interstellar obscuration of light. Such tests are limited to stars found in clusters or galaxies whose distances can be determined by other methods. For this reason, the period-luminosity relation has been regarded by many astronomers as a tool for measurement rather than a precise test of the theory of pulsation.

But there are other properties that can be derived without knowing the star's distance, and these can reveal something about the interior of pulsating stars. (It is much the same when we judge from the high pitch of a bell that it is small, without paying attention to its loudness, which will be affected by distance.) One class of these methods involves measuring the periods of the overtone oscillations; another depends on the shapes of the curves of variation. These properties are sensitive to the structure of the outer layers of the star and can be determined without knowing anything about the stars' distances.

In 1926, Eijnar Hertzsprung,[10] pointed out that the shapes of the light curves of classical Cepheids show a progression with period (also see refs. 10, 14). Stars with periods in the range 6–10 days tend to have a bump on the descending branch of the light curve and stars with periods of 10–17 days have a bump on the ascending branch. Since then, the same pattern has been found among the velocity curves. In fact, the two types of curves show similar shapes.

Each star is capable of a variety of overtones (modes). We will use P_0, P_1, P_2, etc., to indicate the periods of the fundamental, first overtone, second overtone, etc. If a star has evolved to the region of the classical Cepheids, its fundamental mode becomes unstable; it is self-excited, and the star oscillates with a period P_0.

Norman Simon[21,22,23] has shown that the Hertzsprung progression can probably be explained by a resonance between the fundamental and the second overtone, when the period of the overtone is exactly twice the fundamental period. (Also see refs. 19, 20.)

Thus, if the models show a $P_0/P_2 = 2:1$ resonance for stars with $P_0 = 10$ days, then the Hertzsprung progress can probably be understood. Indeed, by adjusting the masses of the models, it was possible to obtain the $P_0/P_2 = 2:1$ resonance. But, until recently, there was a difficulty, because the value of stellar mass that was required to produce the resonance for 10-day Cepheids was nearly twice the value determined from the evolutionary tracks on the HR diagram. Simon noticed that the model's predictions about the Hertzsprung progression were very sensitive to the radiative opacity of stellar material, and he urged that the new calculations be undertaken[21,22,23]. In fact, he was able to predict the amount of error that would be found when the new opacities were calculated, and his predictions were confirmed by the new calculations. (See, for example, refs. 1, 12, 13, 16.)

7.8 Exercises

Although the layering of the model is crude (due to the small number of shells that can be handled), the algorithms are based on current research techniques, and the program can be used to survey and illustrate several areas of current research.

7.1 Constant-Density Model

Carry out the integration of Eq. 7.12 for $\rho = constant$ and find the expression for the pressure, adjusting the constant of integration so that pressure vanishes at the surface.

7.2 Linear Density

Now let the density be a linear function of radius $\rho(r) = \rho_{center}(1 - r/R)$ and find the mass $M(r)$ and pressure $P(r)$.

The following exercises are to be carried out with adiabatic pulsations.

7.3 Pulsation Periods for the Sun

a. With the **EnterParameters** window, set the M, R, L of a model equal to unity in solar units. (Keep the default value of chemical composition.) Build the four types of models and note their fundamental periods, P_0. They differ because the density structure is different. The homogeneous model has the longest period. Can you think of a reason for this?

b. Experiment with different numbers of shells to see how the accuracy degrades when there aren't enough shells.

7.4 Period-Density Relation

a. Use the program to construct 16 homogeneous models with the following masses $M = [0.33, 1.0, 3.0, 10.0]$ and the following R, L pairs: $[1, 1]$, $[2, 4]$, $[3, 9]$, $[10, 100]$. Find their fundamental periods.

b. What do you notice about the surface temperatures of these models? Why is that the case?

c. Devise a program or use a spread-sheet program to compute the mean densities in solar units (M/R^3) for each model. Make a log-log plot of fundamental period against the the mean densities, and test the approximate relation $P\rho^{\frac{1}{2}} = constant$.

d. Do the same for the models in radiative + convective equilibrium. How much of a shift is there between the relationships for the two types of models?

7.5 Modal Shapes

a. Open the **ModePhasors** window and the **Perturbations** window. Start the motion in the fundamental mode and compare the shapes of the perturbations for different variables.

b. Now activate different modes and see how the shapes change.

c. Carry out a and b using the **Strip Chart** window.

7.6 Phase Diagram

a. Open the **PhaseDiagram** window and select velocity and displacement. Activate the fundamental mode. Can you see why the trajectories go clockwise? What will happen when you activate the first overtone? Try it.

b. Try different pairs of variables and make a list of variables that change in phase with each other (such as density and pressure) and those that are out of phase (such a velocity and density).

c. Are these relationships maintained in all the overtones?

7.7 Evolutionary Period Changes
With the stellar evolution simulation of Richard Kouzes (chapter 5) estimate how rapidly the radius of a star increases as it evolves across the HR diagram from the main sequence to the region of giants. Use the radii corresponding to two models separated by 100,000 years and evaluate the difference of period. If the time of maximum of a variable can be determined to within 0.05 days, how many years' observations would be needed to notice the evolutionary period change?

7.8 Mode Resonances

a. Construct a model Cepheid with $M = 2.5M_{sun}$, $R = 45R_{sun}$, and $L = 2500L$. Find the period ratios for the overtones.

b. What happens to the period ratios vary as you increase the radius of the model? Can you find a model that gives $P_0/P_2 = 2$?

7.9 Hertzsprung Progression
By adjusting the phases and amplitudes of the fundamental and second overtone, see if you can generate a progression of shapes of the velocity curve. Use the **Strip Chart** window to view the pulsation.

7.9 Suggested Modifications of the Program

1. **Effect of Opacity**
   Try multiplying the opacity by a factor 2 and see how it affects the periods of Cepheids.

2. **Chemical Inhomogeneity**
   Alter the integration of the equation of hydrostatic equilibrium to take care of the case in which the abundance of helium, Y, is a function of M.

3. **Constant of Gravitation**
   What will happen to the fundamental period of a star if you double the constant of gravitation G? Make a model, note its period, and then go to the unit COMMON.PAS and make the alteration. Construct a model to test your guess.

7.10 Acknowledgments

Irwin Shapiro, Director of the Harvard-Smithsonian Center for Astrophysics, supported this work, and Philip Sadler, Director of its Science Education Department, provided the environment in which it was performed.

References

1. Andreasen, G. K. Stellar consequences of enhanced metal opacity. Astronomy and Astrophysics **201:**72–79, 1988.

2. Baker, N. Simplified model of Cepheid instability. In *Stellar Evolution*, eds. R. F. Stein, A. G. W. Cameron, pp. 333–346 New York: Plenum Press, 1966.

3. Buchler, J. R., Moskalik, P., Kovacs, G. A survey of bump Cepheid model pulsations. Astrophysical Journal **351:**617–631, 1990.

4. Castor, J. I. On the calculation of linear, nonadiabatic pulsations of stellar models. Astrophysical Journal **166:**109–129, 1971.

5. Christy, R. F. The calculation of stellar pulsation. Reviews of Modern Physics **36:**555–571, 1964.

6. Cox, J. P., Giuli, R. T. *Principles of Stellar Structure*. New York: Gordon and Breach, 1968.

7. Cox, J. P. *Theory of Stellar Pulsation*. Princeton, NJ: Princeton University Press, 1980. Contains a comprehensive bibliography and historical perspective.

8. Eddington, A. S. *The Internal Constitution of the Stars*. Cambridge: Cambridge University Press, 1930.

9. Eddington, A. S. On the Cause of Cepheid Pulsation. Monthly Notices of the Royal Astronomical Society **101:**182–194. The original statement of theory of instability as a heat-engine involving ionization in the outer envelope. Later revised by others to include helium, but the ideas are essential unchanged.

10. Hertzsprung, E. On the relation between period and form of the lightcurve of variable stars of the d Cephei type. Bulletin of the Astronomy Institute of the Netherlands III; No. 96, 115–120, 1927.

11. Iglesias, C. A., Rogers, F. J., Wilson, B. G. Opacities for classical Cepheid models. Astrophysical Journal **360:**221–226, 1990.

12. Payne-Gaposchkin, C., Gaposchkin, S. Relation of Light Curve to Period for Cepheids in the Small Magellanic Cloud. Vistas in Astronomy **8:**191–201, 1966.

13. Kovacs, G., Kisvarsanyi, E. G., Buchler, J. R. Cepheid radial velocity curves revisited. Astrophysical Journal **351**:606–616, 1990.

14. Kovacs, G., Buchler, J. R., Marom, A. RR Lyrae pulsations revisited. Astronomy and Astrophysics **252**:L27–L30, 1991.

15. Moskalik, P., Buchler, J. R., Marom, A. Toward a resolution of the bump and beat Cepheid mass discrepancies. Astrophysical Journal **385**:685–693, 1992.

16. Press, W., Flannery, B. P., Teukolsky, S. A., Vetterling, W. T. *Numerical Recipes: The Art of Scientific Computing*. New York: Cambridge University Press, 1986.

17. Rogers, F. J., Iglesias, C. A. Radiative atomic Rosseland mean opacity tables. Astrophysical Journal Supplement Ser. 79, no. 2, 507–568, 1992.

18. Rosseland, S. *The Pulsation Theory of Variable Stars*. Oxford: Clarendon Press, 1949.

19. Schwarzschild, M., Savedoff, M. Anharmonic pulsations of the standard model. Astrophysical Journal **109**:298, 1949.

20. Seya, K., Tanaka, Y., Takeuti, M. Nonlinear two- mode coupling models of radial stellar oscillations. Publication Astronomical Society of Japan, **42**:405–418, 1990.

21. Simon, N., Schmidt, E. G. Evidence favoring nonevolutionary Cepheid masses. Astrophysical Journal **205**:162–164, 1976. Contains references to earlier work on the Hertzsprung progression.

22. Simon, N. Resonance effects and the Cepheid 'bump mass' anomaly. Astrophysical Journal **217**:160–170, 1977.

23. Simon, N. A plea for reexamining heavy element opacities in stars. Astrophysical Journal **260**:L87–L89, 1982.

24. Stellingwerf, R. Modal stability of RR Lyrae stars. Astrophysical Journal **195**:465–466 (erratum: Astrophysical Journal **199**:705), 1975.

25. Stobie, R. S. Cepheid pulsation—I. Monthly Notices of the Royal Astronomical Society 144, No. **4**:461–484, 1969.

26. Zhevakin, S. A. Physical basis of the pulsation theory of variable stars. Annual Review of Astronomy and Astrophysics **1**:367–400, 1963. Review of his earlier work modifying Eddington's theory and demonstrating the importance of helium.

8

Model of a Stellar Atmosphere

Charles A. Whitney

8.1 Introduction

8.1.1 The Role of Model Atmospheres

The atmosphere of a star contains all layers which may be seen from the free space outside the star. Heat radiation produced in the interior of a star must pass through the atmosphere, and the gases of the atmosphere filter the radiation and add features that can be used to determine the temperatures and pressures in the visible layers.

Models of stellar atmospheres have held the key to many problems in stellar astronomy. They are the link between the astronomer and the hidden interiors of stars. They are central to studies of chemical composition of the universe, accretion discs and star formation, close binary stars, stellar winds, and mass-loss from stars.

Many of these problems are beyond the scope of the simplified code in this simulation, but the simulation will introduce you to an important task of modern astronomy: the interpretation of stellar brightness and color in terms of mass, radius, intrinsic brightness (luminosity), and distance. The program will introduce you to the brightness-color diagram, which is also known under the general title of Hertzsprung-Russell, or H-R, diagram. The H-R diagram is undoubtedly the most useful diagram in stellar astronomy.

8.1.2 Two Diagrams for Representing Stellar Models

The first step in building a model for an observed star is to locate it on the H-R diagram. This can be done by determining its intrinsic brightness, that is, the brightness it would have if it were at a standard distance of 10 parsecs. (One parsec = 3.26 light years, the distance light travels in a vacuum in one year. We will use light years in this chapter.) We also need to determine the surface temperature of the star.

There are many ways to measure the brightness and temperature of a star, and the different methods produce different types of H-R diagrams. In all cases, brightness is plotted vertically on a logarithmic scale, and a measure of the temperature is plotted horizontally, with cooler stars to the right. The observational measure of temperature is the color of the star as determined from photometry at several wavelengths. The H-R diagrams in chapters 5 and 6 are plotted in a way that is preferred by theoreticians. The vertical axis is based on total luminosity integrated over the entire spectrum, including radiation that is invisible from the Earth's surface. The horizontal axis is the logarithm of the surface temperature.

Because the coordinates of this theoretical diagram cannot be directly measured from Earth, we must transform those coordinates to empirical measures. This transformation is one of the principal uses of model atmospheres.

In this chapter, we shall use a brightness-color diagram in which each star is represented by a point whose vertical location is determined by the star's absolute brightness. This is derived from the apparent brightness with a correction to move the star to the standard distance of 32.6 light years. The color plotted along the horizontal axis is determined by the logarithm of the ratio of spectral intensity in two carefully specified wavelength bands: the V and B bands, whose letters indicate *visual* and *blue*. This quantity, multiplied by 2.5 to change it to magnitudes, is known as the B-V color of the star. (The V and B bands correspond to the yellow-green and the blue regions of the spectrum. Measurement in the V band corresponds quite closely to the human eye.)

8.1.3 What You Can Do With This Program

The program permits you to select a constellation, point to a star, and see it plotted on a brightness-color diagram. Your task is to build a model atmosphere that imitates the real star, and you do this by specifying numerical values for the basic stellar parameters: radius, mass, and luminosity. The program then builds the model and displays it on the brightness-color diagram, and you may also plot its detailed spectrum.

In this way, you will see how astronomers are able to infer stellar dimensions from close inspection of stellar spectra.

This program does not build a complete, self-consistent stellar interior; that is the task of programs in chapter 5 and chapter 6 by Richard Kouzes. It merely accepts the surface parameters (M, R, L) you specify and computes the observational consequences. If you wish to see how your star fits into the scheme of stellar evolution, you might review the earlier chapters.

8.2 Building a Model

8.2.1 Atmospheric Conditions Assumed

Our sun is the only star whose surface may be mapped in detail, although studies of eclipsing binaries and interferometric studies of a few nearby stars have revealed gross features. The maps are very complicated and they show disturbances on all observable scales. Our simplified approach will ignore the fluctuations and assume that the gas is stratified in homogeneous shells concentric with the center of the star. Further, we assume the shells to be very thin, so the atmosphere may be represented by plane parallel layers. We also assume that the gas ionization obeys the Boltzmann-Saha law, and that the ratio of absorption to emission is given by the Planck function.

The simulation permits you to exercise several options about the calculation of opacity and the thermal structure of the atmosphere. You may select to build a model in which thermal energy is brought from the interior by radiation alone. In such models, the temperature gradient can become very steep; in fact, it can become so steep that the density actually decreases inward. In a real star, this would imply that the layers have become unstable and will begin to turn over and undergo convection, like the Earth's atmosphere near the ground when it is heated by direct sunlight. In this case the model would be highly unrealistic, so you are given the option of including thermal convection in the heat transfer. A mixing-length theory of convection is used. (For general discussions of model atmospheres, see refs. 2, 3, 4, 6, 7.)

8.2.2 Conservation Laws Governing an Atmosphere

The photons generated in the deep interior are trapped by overlying gas, while photons generated in the atmosphere may escape to space. The transition between these two regions is not sharp, and several of the equations used in the interior may be used in the atmosphere. But the fact that some of the photons may escape the atmosphere has a profound effect on our method of model building and, in particular, our treatment of the energy transport. (As you can see by looking at a textbook on stellar atmospheres, the mathematical treatment of radiation in an atmosphere is a highly developed branch of mathematical astrophysics. Fortunately, we can make good progress with a highly simplified treatment.)

Hydrostatic Equilibrium

For most stars, we may assume the atmosphere to be thin and to contain negligible mass. We replace the radial distance by height, z, and write Eqs. 8.1 and 8.2 as

$$M(z) = M = constant, \tag{8.1}$$

and for the equation of hydrostatic equilibrium we have

$$\frac{dP(z)}{dz} = -g\rho(z), \tag{8.2}$$

where the surface gravity is $g = GM/R^2$.

Boundary Conditions

As we move outward, the pressure decreases to a value that is much smaller than the pressure in the spectrum-forming region. In order to integrate Eq. 8.2 we must have a starting point. It is not enough merely to specify vanishing pressure, because this would imply vanishing density. By Eq. 8.2 we would then have a vanishing pressure gradient, and the integration would get nowhere.

In fact, specifying a satisfactory starting point is one of the knottiest aspects of model atmosphere theory and we are forced to take a highly simplified approach. Fortunately, the significant layers, deeper down, are not much affected by our approximation. (In fact, most errors tend to self-cancel as we work downward.)

Because the density is very low in the upper layers, thermal energy in a stationary atmosphere can be effectively transported only by radiation and thermal conduction. At such low densities, convection is unimportant, and the happy result is that we may treat the outermost layers as though they were at a constant temperature. The value of the temperature is determined by the radiative flux through the atmosphere, as we shall see.

Nuclear energy sources are negligible in stellar atmospheres, so when energy transport is entirely by heat radiation, we can write the conservation of energy as

$$L(z) = L = constant. \tag{8.3}$$

For an atmosphere, the flux per unit area, F, is more relevant, where we define

$$F = \frac{L}{4\pi R^2}. \tag{8.4}$$

The three global parameters, M, L, and R thus translate into two local parameters, g and F, when we specify an atmosphere. These local parameters may vary with height according to an externally specified relationship, but we will assume g to be constant. (In treating accretion discs, for example, it is assumed that the disc is self-gravitating and gravity increases with height according to $g = az$.)

Equation of State and Opacity

In the conditions of stellar atmospheres, the ideal gas law applies, so we may compute the gas pressure from the particle density, N, the temperature, T, and the Boltzmann constant, k,

$$P = NkT. \tag{8.5}$$

The chemical composition of the gas is defined in terms of the fractional abundances of hydrogen, helium, and metals, by number, X, Y, and Z, as described in chapter 5. (The "metals" include all elements heavier than helium.)

The particle density is the sum of particle densities for individual species:

$$N = N(\text{protons}) + N(\text{hydrogen atoms})$$
$$+ N(\text{helium atoms}) + N(\text{helium ions})$$
$$+ N(\text{metal atoms}) + N(\text{metal ions})$$
$$+ N(\text{electrons}) + N(\text{negative hydrogen ions}).$$

Fractional ionizations, such as $N(\text{protons})/(N(\text{protons}) + N(\text{hydrogen atoms}))$ are computed from temperature and electron number density, using the Saha-Boltzmann ionization law. Further details were provided in chapter 5.

The characteristic opacity $\kappa_0(\rho, T, Y, Z)$, which we will need for solving the equation of hydrostatic equilibrium, is the Rosseland mean computed from a polynomial approximation.

Monochromatic opacities for hydrogen and its negative ion, and for helium are taken from Mihalas's review article[7].

Energy Conservation

Radiation is crucial to the energy balance of an atmosphere. Unfortunately, the literature contains more than one definition of radiant flux. The three most important are

$$F_v(z) = \text{the energy flow per unit area per frequency interval} \qquad (8.6)$$

$$\mathscr{F}_v(z) = \frac{F_v}{\pi,} \text{ the astrophysical flux} \qquad (8.7)$$

$$H_v = \frac{F_v}{4\pi,} \text{ the flux per steradian.} \qquad (8.8)$$

The last is called the Eddington flux, in honor of the great astrophysicist, Sir Arthur Eddington.

The total flow of energy carried by radiation is an integral over frequencies, and we indicate this by dropping the frequency subscript:

$$F = \int_0^\infty F_v \, dv. \qquad (8.9)$$

The condition of energy conservation can be expressed in terms of the luminosity, $L(r)$, which is the total energy outflow across a sphere of radius r centered on the star. The "effective" temperature T_e of the star is the temperature of a blackbody surface that would emit the same energy. It is the temperature at the effective depth of photon formation and for a gray atmosphere it is about 20 percent greater than the temperature at the top. In summary

$$F = \frac{L}{4\pi R^2} = \pi\mathscr{F} = 4\pi H = \sigma_B T_e^4, \qquad (8.10)$$

where $\sigma_B = 5.67 \times 10^{-5}$ (erg cm^{-2} sec$^{-1}T^{-4}$) is the Stefan-Boltzmann constant.

The energy may be carried in a variety of modes, and for stellar atmospheres it is adequate to indicate the flux as the sum of four terms:

$$\text{Total Flux} = F_r + F_{conv} + F_{m-a} + F_{cond},$$

where F_r is the radiation flux; F_{conv} is the convective flux, which is important in the deeper layers; F_{m-a} is the magneto-acoustic flux carried by waves generated below the atmosphere and propagated upward; and F_{cond} is thermal conduction, primarily by electrons.

The first three terms are important in most stellar atmospheres, to varying degrees depending on the underlying star and the location within the atmosphere. The theories for transport by convection and wave dissipation are still very crude and we will not describe them, although the simulation program uses a mixing-length theory for convection. Thermal conduction is fairly well understood, and it is important high above the atmospheric layers where most of the spectrum is formed. We shall ignore it, as it probably has no effect on the theory of spectral types.

8.2.3 Radiation Transfer in an Atmosphere

In the discussion of stellar interiors in chapter 4, the flow of photons was treated as a diffusion process because the photon mean free path is very short compared to distances in which the temperature changes significantly. To put the matter in slightly different terms, the physical parameters at a height, z, are all determined by conditions in the immediate neighborhood, that is, by values of $P(z)$, $T(z)$, and the gradient of T. We say that the variables are "locally determined" in a stellar interior.

By contrast, in a stellar atmosphere the photons at a particular point carry energy and information from distant regions. The entire structure is connected by photons, and the conditions are no longer locally determined. (As a loose analogy to this distinction, we might say that before the days of long-distance communication, social and economic conditions in a village were locally determined. The advent of communication and transport of goods made each village dependent on distant villages.)

For this reason, the mathematical problems of model atmospheres are quite different from those of model interiors. In the interior, the equations are solved by numerical integrations from the surface and from the center. Successive trials (iterations) are required in order to find the right starting conditions at each end of the interval.

In the atmosphere, the need for iterations is even more acute. We cannot integrate the equation of hydrostatic equilibrium until we know the temperature, and we cannot know the temperature until we have found the radiation field from the opacity and the emission at all points in the atmosphere

The way out of this deadlock is by iteration. We adopt an approximate temperature distribution, integrate the equation of hydrostatic equilibrium, and then find the radiation field. If the flux of energy varies by more than a few percent, indicating a violation of energy conservation, we adjust the temperature, recompute the model, and repeat the process.

Differential Equation for Beam Intensity With Pure Absorption

Consider a collimated beam of photons with frequencies in the range $(v, v + dv)$ moving along the direction $\vec{s}$. Construct a small surface element $d\sigma$ perpendicular to $\vec{s}$. If $d\sigma$ is sufficiently small, the amount of energy carried across $d\sigma$ by the beam per unit time will be proportional to the product $dv\,d\sigma$ and we may write it as $I_v\,dv\,d\sigma$, where I_v is the beam intensity. It is the energy flow per unit area per unit frequency and per unit time within the beam. By assuming the radiation is collimated along two beams, and using the concept of beam intensity rather than the conventional specific intensity, we avoid some of the mathematical details that will be found in standard texts.

If, now, we move a short distance along $\vec{s}$ and construct another surface $d\sigma$, the energy flowing across this surface will differ due to losses and gains along the way. We write

$$(I'_v - I_v)\,dv\,d\sigma = -\text{energy lost} + \text{energy gained}. \qquad (8.11)$$

In a vacuum, no energy will be gained or lost. In the presence of matter of density ρ, we assume that the amounts lost and gained will be proportional to $\rho\,ds$. Further, we assume the amount lost is proportional to the incident flux. We introduce two factors:

$$\kappa_v \equiv \text{mass absorption coefficient,}$$

$$j_v \equiv \textit{mass emission coefficient,}$$

which are defined so

$$Energy\ lost = I_v \kappa_v \rho\ dv\ d\sigma\ ds,$$

$$Energy\ gained = j_v \rho\ dv\ d\sigma\ ds.$$

The net change is

$$(I'_v - I_v)\ dv\ d\sigma = (-I_v \kappa_v + j_v)\rho\ dv\ d\sigma\ ds. \tag{8.12}$$

Dividing by $dv\ ds\ d\sigma$, we find the differential equation for the beam intensity:

$$\frac{dI_v}{ds} = -\kappa_v \rho I_v + j_v \rho. \tag{8.13}$$

In the absence of emission (say, in a cold gas)

$$dI_v = -I_v \kappa_v \rho\ ds, \tag{8.14}$$

and if we define optical depth by

$$d\tau'_v = \kappa_v \rho\ ds, \tag{8.15}$$

we have $d\ln(I_v) = -d\tau'_v$, which integrates to

$$I_v(\tau'_v) = I_v(0)\ \exp(-\tau'_v). \tag{8.16}$$

In an equilibrium thermal enclosure (such as a furnace) the gradient of I_v vanishes so $I_v = j_v/\kappa_v$. We also know that the beam intensity is equal to the Planck function, $I_v = B_v$. Thus we have

$$j_v = \kappa_v B_v(T). \tag{8.17}$$

This is the Kirchhoff-Planck relation, valid in a thermal enclosure. We shall use it to find the emission from the opacity of the gas in an atmosphere. (This is usually called the assumption of "local thermodynamic equilibrium." The validity of this assumption is briefly discussed in section 8.3.3.)

In the atmosphere we will measure optical depth along the vertical, increasing downward, so

$$d\tau_v = -\kappa_v \rho\ dz. \tag{8.18}$$

Imagine that all the radiation flows along two anti-parallel beams tilted at an angle Θ with the vertical. This is the "two-stream" approximation which we will use throughout this discussion, and we adopt the convention $0 < \Theta \leq 90^o$ and write $\mu = \cos(\Theta)$.

For the upward and downward beams, we have

$$\mu \frac{dI^+_v}{d\tau_v} = I^+_v - S_v, \tag{8.19}$$

$$-\mu \frac{dI^-_v}{d\tau_v} = I^-_v - S_v, \tag{8.20}$$

where we have defined the "source function for pure absorption," S, by

$$S_v \equiv \frac{j_v}{\kappa_v}. \tag{8.21}$$

The net flux across a horizontal surface is

$$H_v = \int \mu I_v \frac{d\omega}{4\pi} = \frac{\mu}{2}(I_v^+ - I_v^-), \tag{8.22}$$

and adding the two differential Eqs. 8.19 and 8.20, we find

$$\frac{dH_v}{d\tau_v} = J_v - S_v, \tag{8.23}$$

where we have defined the mean intensity, J_v, by

$$J_v \equiv \int I_v \frac{d\omega}{4\pi} = \frac{1}{2}(I_v^+ + I_v^-). \tag{8.24}$$

Subtracting equation Eq. 8.20 from Eq. 8.19 and using Eq. 8.23, we find

$$\frac{dJ_v}{d\tau_v} = \frac{H_v}{\mu^2}. \tag{8.25}$$

One further manipulation is useful. Differentiating Eq. 8.25, we find

$$\frac{d^2 J_v}{d\tau_v^2} = \frac{1}{\mu^2}\frac{dH_v}{d\tau_v} = \frac{1}{\mu^2}(J_v - S_v). \tag{8.26}$$

This relation forms the basis of the Feutrier method for finding J from S, described later.

Differential Equation for Beam Intensity With Scattering and Absorption

We must add a complication, because photons can be scattered by free electrons as well as being absorbed by atoms. If this scattering were the only interaction with the photon beams, the differential equation for the beam intensities would be

$$\mu\frac{dI_v^+}{dz} = -\sigma_e \rho I_v^+ + \frac{1}{2}\sigma_e \rho(I_v^+ + I_v^-), \tag{8.27}$$

$$-\mu\frac{dI_v^-}{dz} = -\sigma_e \rho I_v^- + \frac{1}{2}\sigma_e \rho(I_v^+ + I_v^-). \tag{8.28}$$

We let σ_e be the Thompson scattering coefficient per gram. We assume it is independent of frequency, and we neglect the small Compton shift of frequency produced in each scattering. Further, we assume half the photons are scattered forward into the same beam and half are scattered backward into the other beam. Adding the effects of absorption, we find the full transfer equation for each beam,

$$\mu\frac{dI_v^+}{dz} = -(\sigma_e + \kappa_v)\rho I_v^+ + \sigma_e \rho J_v + \kappa_v \rho B_v, \tag{8.29}$$

$$-\mu\frac{dI_v^-}{dz} = -(\sigma_e + \kappa_v)\rho I_v^- + \sigma_e \rho J_v + \kappa_v \rho B_v. \tag{8.30}$$

Introducing the total optical depth

$$d\tau_v = -(\kappa_v + \sigma_e)\rho\, dz, \tag{8.31}$$

and the source function

$$S_v \equiv \frac{\sigma_e J_v + \kappa_v B_v}{\sigma_e + \kappa_v},$$

(8.32)

we find forms identical to Eqs. 8.19 and 8.20. Those are the fundamental equations. The only effects of adding scattering are a change in the calculation of optical depth and the source function, which becomes an explicit function of the local radiation as well as the temperature.

It is convenient to introduce the fractional absorption

$$\epsilon_v \equiv \frac{\kappa_v}{\kappa_v + \sigma_e},$$

(8.33)

so

$$S_v = \epsilon_v B_v + (1 - \epsilon_v) J_v.$$

(8.34)

Analytic Solution for Gray Atmosphere

If we assume that opacity is independent of frequency (not realistic but very helpful), the atmosphere is said to be gray. We may integrate Eqs. 8.19–8.34 over frequency and drop the v subscript. This simplification makes the equations solvable in closed form. For a radiative equilibrium atmosphere we have, from Eq. 8.25,

$$\frac{dJ(\tau)}{d\tau} = \frac{H}{\mu^2} = constant,$$

(8.35)

which integrates to

$$J(\tau) = \frac{H}{\mu^2}\tau + J(0).$$

(8.36)

The mean intensity increases linearly with optical depth. To complete the solution we need the constant of integration, $J(0)$. To derive it, we use the conditions at the surface:

$$H = \frac{\mu I^+(0)}{2}, \qquad J(0) = \frac{I^+(0)}{2}.$$

(8.37)

From these we have

$$J(0) = \frac{H}{\mu},$$

(8.38)

so the solution is

$$J(\tau) = \frac{H}{\mu^2}(\tau + \mu).$$

(8.39)

To find the temperature from the mean intensity, we use the the radiative equilibrium condition $H = constant$. With Eq. 8.23 this gives $J = S$. Inserting the expression for S (Eq. 8.34), we find

$$J(\tau) = \epsilon B(\tau) + (1 - \epsilon) J(\tau).$$

(8.40)

This gives

$$J(\tau) = B(\tau) = \frac{\sigma}{\pi} T^4(\tau), \tag{8.41}$$

$$T^4(\tau) = \frac{\pi}{\sigma} \frac{H}{\mu^2}(\tau + \mu). \tag{8.42}$$

This may be written

$$T^4(\tau) = T_e^4 \frac{(\tau + \mu)}{(4\mu^2)}, \tag{8.43}$$

so

$$\frac{dT}{d\tau} = T \left(\frac{T_e}{T}\right)^4 \frac{1}{(16\mu^2)}. \tag{8.44}$$

Notice that ϵ does not appear in the expression for the temperature distribution. This means that scattering does not affect the temperature in a gray radiative equilibrium atmosphere. (This is so because the scattering events are assumed not to affect the thermal energies of the electrons.) However, we shall see that scattering does affect the emitted spectrum, as it increases the effective depth of photon formation.

If $H(\tau)$ is a known function of optical depth, Eqs. 8.23 and 8.25 with the boundary condition $J(0) = H(0)/\mu$ give the following expression for the corresponding source function:

$$S(\tau) = \frac{1}{\mu^2} \int_0^\tau H\, d\tau - \frac{dH}{d\tau} + H(0). \tag{8.45}$$

From this we may derive an expression that is useful for improving a model to make its flux more nearly constant. Define $S^0(\tau)$ as the source function for constant flux, H^0, so

$$S^0(\tau) = \frac{H^0}{\mu^2}(\tau + \mu). \tag{8.46}$$

Now write the variable flux as a constant plus a depth-dependent term, $H = H^0 + \delta H$, and write the corresponding source function as $S = S^0 + \delta S$. Inserting these into Eq. 8.45 we easily find, after a little cancellation,

$$\delta S = \frac{1}{\mu^2} \int_0^\tau \delta H\, d\tau - \frac{d}{d\tau} \delta H + \frac{1}{\mu} \delta H(0). \tag{8.47}$$

This equation permits us to derive the change of source function corresponding to a prescribed change of flux. In the case of pure absorption, $\epsilon = 0$, so $S = B$, and the corresponding change of temperature is given by

$$\frac{\delta T}{T} = \frac{1}{4} \frac{\delta S}{S}. \tag{8.48}$$

8.3 Computational Approach

The construction of a model by iteration consists of two types of calculations. First we integrate the equation of hydrostatic equilibrium from the surface inward with an assumed temperature, then we evaluate the radiative flux. If the calculated flux does not follow the prescribed flux (for example, it if varies when we want it to be constant), we alter the temperature and repeat the process. (The simulation permits you to make manual changes in the temperature distribution.)

8.3.1   Hydrostatic Equilibrium

With Eq. 8.18 we may rewrite Eq. 8.2 to find

$$\frac{dP}{d\tau} = \frac{g}{\kappa_0(\rho, T)}. \tag{8.49}$$

To integrate this equation, we set up a grid of optical depth points, $\tau_i(i = 1, 2 \ldots N)$ and find an approximate value of pressure at the first point. Numerical integration by a Runge-Kutta scheme takes us down into the atmosphere and we stop at a comfortably deep layer, say, $\tau = 10$.

8.3.2   Methods for Radiation

In evaluating the radiation in any stage of the iteration, we adopt the gas structure from the previous iteration. At each optical depth we know the temperature and therefore B_v and ϵ_v, but we need the source function defined by Eq. 8.32. This has contributions from thermal emission and scattered radiation. Since J depends on S throughout the atmosphere, we must solve for radiation in the whole atmosphere at once.

Using the grid of N optical depth points, we establish a set of N simultaneous linear equations for $J_v(\tau_i)$ at each frequency. The sets are independent of each other, and they are solved frequency-by-frequency for $J_v(\tau_i)$ and $S_v(\tau_i)$.

In order to discuss the non-gray case later, we define the monochromatic opacity ratio,

$$\chi_v \equiv \frac{d\tau_v}{d\tau_0}, \tag{8.50}$$

in terms of the reference optical depth τ_0. Values for $\chi_v(\rho, T)$ are computed from the expression in ref. 6. We may write Eqs. 8.23 and 8.25 in terms of the representative optical depth, as

$$\frac{1}{\chi_v} \frac{dH_v}{d\tau} = J_v - S_v \tag{8.51}$$

and

$$\frac{\mu}{\chi_v} \frac{dJ_v}{d\tau} = \frac{H_v}{\mu}. \tag{8.52}$$

Inserting the second of these into the first, we find

$$\frac{\mu^2}{\chi_v} \frac{d}{d\tau} \left(\frac{1}{\chi_v} \frac{dJ_v}{d\tau} \right) = J_v - S_v. \tag{8.53}$$

To simplify notation, the frequency subscript will be dropped with the understanding that all quantities are monochromatic in the remainder of this section. So we have

$$\frac{\mu}{\chi} \frac{dJ}{d\tau} = \frac{H}{\mu}, \tag{8.54}$$

$$\frac{\mu^2}{\chi} \frac{d}{d\tau} \left(\frac{1}{\chi} \frac{dJ}{d\tau} \right) = J - S. \tag{8.55}$$

This is the form we will solve with the Feutrier finite-difference formulation.

Boundary Conditions

In order to solve this second-order differential equation, we must express the conditions that hold at the boundaries of the atmosphere. Specifically, we need relations between the mean intensity at each boundary and the value just inside that boundary.

Auer[1] has shown that, for reliable computations, we must use a treatment that is accurate to second-order in optical depth near the surface. In what follows, all quantities are monochromatic and τ is the reference optical depth. At the surface of the star the radiation is confined to the upward direction, so

$$I^- = 0, \qquad J(0) = \frac{I^+(0)}{2}, \qquad H(0) = \mu J(0). \tag{8.56}$$

We first express $J(\delta\tau)$ in terms of a Taylor series about the surface,

$$J(\delta\tau) = J(0) + \frac{dJ}{d\tau} \delta\tau + \frac{1}{2} \frac{d^2J}{d\tau^2} \delta\tau^2. \tag{8.57}$$

The coefficient of the first-order term is evaluated from

$$\frac{dJ}{d\tau} = \frac{\chi H}{\mu^2}, \tag{8.58}$$

which gives

$$\frac{dJ}{d\tau} = \frac{\chi J(0)}{\mu} \tag{8.59}$$

at the boundary. The coefficient of the second-order term is evaluated by exploiting the second-order differential equation evaluated at the surface with $\chi = constant$,

$$\frac{1}{2\chi^2} \frac{d^2J}{d\tau^2} = \frac{1}{2\mu^2}(J - S). \tag{8.60}$$

Using

$$S = \epsilon B + (1 - \epsilon)J, \tag{8.61}$$

we find the desired expression for $J(\delta\tau)$ in terms of its surface value and the known value of B at the surface:

$$J(\delta\tau) = J(0) + J(0)\frac{\chi}{\mu} \delta\tau + \frac{\epsilon\chi^2}{2\mu^2}(J(0) - B(0)) \delta\tau^2. \tag{8.62}$$

Deep in the star, the radiation field is approximated by the Planck function. It is nearly isotropic, so a first-order expansion is sufficiently accurate, and we can express the mean intensity just inside the boundary ($i = N$) by

$$\frac{\mu}{\chi} \frac{J_v(\tau) - J_v(\tau - \delta\tau)}{\delta\tau} = \frac{H_v}{\mu} = I_v^+(\tau) - J_v(\tau). \tag{8.63}$$

From the integral of the transfer equation along the outgoing ray we have

$$I_v^+ = B_{vN} + \mu \frac{dB_{vN}}{d\tau}. \tag{8.64}$$

The second term on the right may often be neglected. If it is needed, it may be evaluated from

$$\frac{dB_v}{d\tau} = \frac{dB_v}{dT} \frac{dT}{d\tau}, \tag{8.65}$$

and the value of dT/dt may be taken from the radiative equilibrium solution corresponding to the local radiative flux.

Finite-Difference Form of the Equations

The essence of the Feutrier method for solving the differential equations is to replace them with a set of finite difference equations for the monochromatic mean intensities, $J_i \equiv J(\tau_i)$, ($i = 1, 2 \ldots N$) at a discrete set of optical depths.

We will now show that the equation for the upper boundary condition may be written in the following finite-difference form in terms of a known vector $\mathbf{g}$ (not to be confused with the gravity):

$$B_1 J_1 - C_1 J_2 = g_1, \tag{8.66}$$

while the equation for the general interior layer takes the form

$$-A_i J_{i-1} + B_i J_i - C_i J_{i+1} = g_i, \tag{8.67}$$

and the equation for the bottom boundary condition is

$$-A_N J_{N-1} + B_N J_N = g_N, \tag{8.68}$$

and the equations form a set of inhomogeneous algebraic expressions which may be solved by standard techniques. In matrix form, $\mathcal{T}\mathcal{J} = \mathbf{g}$ and the matrix T is tridiagonal. All terms vanish except the diagonal and adjacent off-diagonal elements. To cast the equations in tridiagonal form, we start with the equation for the surface layers. In the expression for the boundary condition we set $J_1 = J(0)$, $J_2 = J(\delta\tau)$ so

$$J_2 = J_1 + J_1 \frac{\chi}{\mu} \delta\tau + \frac{\epsilon\chi^2}{2\mu^2} \delta\tau^2 (J_1 - B_{v1}). \tag{8.69}$$

or

$$B_1 J_1 - C_1 J_2 = g_1, \tag{8.70}$$

where

$$B_1 = 1 + \frac{\chi}{\mu} \delta\tau + \frac{\epsilon\chi^2}{2\mu^2} \delta\tau^2. \tag{8.71}$$

$$C_1 = 1, \tag{8.72}$$

$$g_1 = \frac{\epsilon \chi^2}{2\mu^2} \delta\tau^2 B_{v1}. \tag{8.73}$$

For the intermediate layers we use

$$\frac{\mu}{\chi} \frac{d}{d\tau}\left(\frac{\mu}{\chi} \frac{dJ}{d\tau}\right) = J - S. \tag{8.74}$$

We define

$$\Delta_i \equiv \frac{(\chi_{i+1} + \chi_i)(\tau_{i+1} - \tau_i)}{2\mu}, \tag{8.75}$$

so the first derivative is

$$\left(\frac{\mu}{\chi} \frac{dJ}{d\tau}\right)_{i+\frac{1}{2}} = \frac{J_{i+1} - J_i}{\Delta_i} \equiv f_{i+\frac{1}{2}} \tag{8.76}$$

and the second derivative is

$$\left(\frac{\mu}{\chi} \frac{df}{d\tau}\right)_i = \frac{f_{i+\frac{1}{2}} - f_{i-\frac{1}{2}}}{\delta_i}, \tag{8.77}$$

where

$$\delta_i \equiv \frac{\Delta_i + \Delta_{i-1}}{2}. \tag{8.78}$$

Introducing the expression for f_i and collecting terms, we have

$$-A_i J_{i-1} + B_i' J_i - C_i J_{i+1} = S_i, \tag{8.79}$$

where

$$A_i = \frac{1}{\Delta_{i-1}\,\delta_i}, \qquad B_i' = 1 + A_i + C_i, \quad C_i = \frac{1}{\Delta_i\,\delta_i}. \tag{8.80}$$

The final step in deriving the difference equation for the intermediate layer is to introduce Eq. 8.61 for the source function and take $(1 - \epsilon_i)J_i$ to the other side. We find

$$-A_i J_{i-1} + B_i J_i - C_i J_{i+1} = g_i, \tag{8.81}$$

where

$$B_i = \epsilon_i + A_i + C_i, \qquad g_i = \epsilon_i B_{vi}. \tag{8.82}$$

The coefficients A, B, C, and g are known. When $\epsilon_i = 1$, the source term on the right becomes the Planck function at the local temperature.

The boundary condition at the bottom of a thick atmosphere may be written

$$-A_N J_{N-1} + B_N J_N = g_N, \tag{8.83}$$

with

$$A_N = \frac{1}{\Delta_{N-1}}, \qquad B_N = 1 + A_N, \qquad g_N = B_{vN} + \mu \frac{dB_{vN}}{d\tau}. \tag{8.84}$$

Solution of the Finite-Difference Equations

We use Gaussian elimination involving two auxiliary variables, D and Z, defined by the pair of recurrence relations (see refs., 6, 8, 9, 10: Procedure Tridiag):

$$D_1 = \frac{C_1}{B_1}, \qquad D_i = \frac{C_i}{B_i - A_i D_{i-1}} \tag{8.85}$$

$$Z_1 = \frac{g_1}{B_1}, \qquad Z_i = \frac{g_i + A_i Z_{i-1}}{B_i - A_i D_{i-1}}, \tag{8.86}$$

from which the solution for J_i follows:

$$J_{N+1} = 0, \qquad J_i = D_i J_{i+1} + Z_i. \tag{8.87}$$

Rybicki and Hummer[9] have rewritten these equations into a form with better numerical conditioning near the surface. They define two auxiliary quantities which we will denote U and V (respectively, F and H in their notation):

$$U_i = D_i^{-1} - 1, \qquad V_i = -A_i + B_i - C_i = \epsilon_i. \tag{8.88}$$

The recursion relations become

$$U_1 = \frac{V_1}{C_1}, \qquad U_i = \frac{V_i + \frac{A_i U_{i-1}}{1 + U_{i-1}}}{C_i}, \tag{8.89}$$

$$Z_i = \frac{g_1}{B_1}, \qquad Z_i = \frac{g_i + A_i Z_{i-1}}{C_i(1 + U_i)}, \tag{8.90}$$

and the solutions are given by

$$J_N = B_{\nu N}, \qquad J_i = \frac{J_{i+1}}{1 + U_i} + Z_i. \tag{8.91}$$

These are the equations used by the simulation program to find the radiation field. (Our equations differ slightly from Rybicki and Hummer[9] because we have introduced the explicit expression for the source function.)

8.3.3 Limitations and Caveats

This chapter is intended as an introduction to radiative transfer theory and to atmospheric modeling. Therefore, we have adopted the simplifying assumption of local thermodynamic equilibrium. This is excellent for the interiors of stars, and it is a useful beginning point, but not an accurate endpoint, for the atmosphere. Given the power of mainframe computers, the attitude of today's researchers in stellar atmospheres is that the Boltzmann distribution of populations among atomic levels is invalid until proven valid. Much of current research is aimed at developing techniques for computing the ionization and emission from more general premises. In many cases, the corrections are significant

The assumption that the atmosphere is stationary represents an equally important limitation on the applicability of this simulation. It eliminates exploring the formation of stellar chromospheres and coronas, features that may play a role in mass-loss and evolution. Observations of the Sun show that significant amounts of energy are transported by sound waves, which dissipate and heat the upper layers. These processes are the focus of most current research on the solar atmosphere. They are beyond the scope of this simulation, although the simulation may be used to study the qualitative relationship between temperature and energy flux in the upper atmosphere.

8.4   Details of the Program

8.4.1   Running the Program

The aim of this program is to find a set of stellar parameters that produce a desired spectrum or put the model star in a desired position on the H-R diagram. You may select a real star for modeling by selecting the menu **See Stars/On Sky** and then choosing a constellation to be plotted. The data for this constellation will be read in from a disk file (see Star Data File, in the next subsection). When you click on one of the stars on the constellation diagram, it will be plotted according to its brightness and color on the H-R diagram to the right. Its photometric properties will be listed at the lower right. (Spectral type is a semi-qualitative classification that is included for completeness only. It is not currently possible to directly compute a spectral type from stellar parameters.)

Depending on the (B-V) color of the star, you will build a relatively hot or cool star, and depending on the absolute visual magnitude, you will specify the star to be large or small. You may also adjust the distance of the model to look for a match to the apparent magnitude of the star. These parameters may be entered by selecting the menu item **Model 1/Specify Model**. Notice that you cannot directly enter the surface temperature or the color of the model. These properties are derived from the mass, radius, and luminosity of the star. Likewise, the surface gravity is derived from the mass and radius.

After you have built a model go back to the H-R diagram and see where the model is plotted. Then adjust the parameters to make this point fall closer to the star you have selected.

You will find, for example, that changing the mass will affect the gravity but it will change the spectrum very little—especially in the gray case, where the gravity has no effect on the spectrum.

In addition to this type of matching game, you may also examine the effect of various temperature distributions on the spectrum and on the depth variation of the radiative flux.

You may make a variety of graphs with the menu **Plot- What**.

See Exercises for other suggestions of things to do.

Star Data File

A special data file must be available if you wish to display constellations and data for real stars. A default file is provided with the program. It contains records for the constellation data and is a special type called "CONS." (You cannot manipulate this file with a text editor.) This file is created by an auxiliary program called CreateStarFile. You may edit the source code of CreateStarFile and add or remove data, then compile and run it. (The format of the code is rather inefficient, but it should be self-explanatory and easy to alter.)

When you run the main STELLAR ATMOSPHERE program and call for a constellation, the program looks to see if a file has already been read in. If not, you will be prompted with a file-opening window which will display the names of files of the correct type. Select the data file you wish and open it. The program will continue to use this file until you exit and restart the program.

Sample Input and Output

As an example, a model of the sun's atmosphere might have the following parameters:

Solar units: Mass = 1; radius = 1; luminosity = 1.
Distance = 32.0 light years.
Surface temperature = 5770K, surface gravity = 2.74×10^4 cm sec^{-2}.
Helium abundance = 0.26, metals abundance = 0.020.
Type of model: Radiative

The following data were derived from this model (see definitions below).

Sun 1

Te	g	M	R	L		
5.770e+3	2.740e+4	1.000e+0	1.000e+0	1.000e+0		
Distance(ly),	mv app,		B-V,	MV(abs),	M(bol)	
32.0		4.79		0.65	4.83	4.75

X	Y	Z	Incl. Rad. Pressure?	MuZero
0.7200	0.260	0.020	FALSE	0.577

Type		Absorber
Radiative		Gray

Number of shells   12

s	Tau	T	P	V	kappa	mmwt
	Height(cm)	adgrad	radgrad	Hion	Heion	He+ion
1	0.010	4701	1.084e+4	2.992e+7	1.068e−1	1.26
	−0.00e+0	0.399	0.018	0.000	0.000	0.000
2	0.019	4718	1.290e+4	2.523e+7	1.252e−1	1.26
	−2.07e+6	0.399	0.025	0.000	0.000	0.000
3	0.035	4750	1.613e+4	2.032e+7	1.528e−1	1.26
	−4.73e+6	0.399	0.037	0.000	0.000	0.000
4	0.066	4809	2.099e+4	1.580e+7	1.922e−1	1.26
	−7.90e+6	0.399	0.057	0.000	0.000	0.000
5	0.123	4913	2.818e+4	1.202e+7	2.457e−1	1.26
	−1.15e+7	0.399	0.090	0.000	0.000	0.000
6	0.231	5091	3.864e+4	9.087e+6	3.187e−1	1.26
	−1.55e+7	0.398	0.139	0.000	0.000	0.000

7	0.433	5383	5.346e+4	6.945e+6	4.349e$-$1	1.26
	$-$1.97e+7	0.397	0.210	0.000	0.000	0.000
8	0.811	5829	7.231e+4	5.561e+6	7.116e$-$1	1.26
	$-$2.40e+7	0.392	0.338	0.000	0.000	0.000
9	1.520	6462	9.055e+4	4.924e+6	1.672e+0	1.26
	$-$2.74e+7	0.375	0.659	0.001	0.000	0.000
10	2.848	7306	1.029e+5	4.906e+6	5.374e+0	1.25
	$-$2.96e+7	0.326	1.474	0.003	0.000	0.000
11	5.337	8375	1.097e+5	5.314e+6	1.904e+1	1.24
	$-$3.09e+7	0.230	3.224	0.012	0.000	0.000
12	10.000	9687	1.134e+5	6.126e+6	6.488e+1	1.20
	$-$3.16e+7	0.141	6.346	0.051	0.000	0.000

These are the variables in the output:

- τ, optical depth of the layer

- $T(K)$, temperature of layer

- $P(d/cm^2)$, pressure (including radiation pressure)

- $V(cm^3/g)$, specific volume, or reciprocal of density

- κ, opacity ($\kappa \rho dx$ is the increment of optical depth, so κ has units cm^2/g)

- μ, mean molecular weight of the gas (dimensionless)

- z, height of the layer (cm)

- $(\gamma - 1)/\gamma$, adiabatic temperature gradient

- $dlnT/dlnP$, actual logarithmic temperature gradient (dimensionless)

- *Hion*, fraction of hydrogen in form of protons

- *Heion*, fraction of helium that is singly ionized

- He^+ion, fraction of helium that is doubly ionized

Figures 8.1–8.3 display typical output.

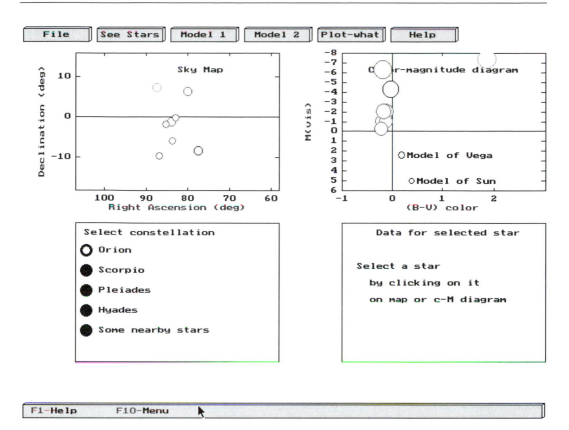

Figure 8.1: This panel permits viewing a selected constellation plotted on the sky (upper left) and plotted in a color-magnitude diagram (upper right). Data for stars selected with the mouse are displayed in the lower right, and one of the challenges is to construct a model star which imitates the real star.

8.5 Exercises

8.1 Effective Temperature

Verify that the effective temperature of the sun is $T_e = 5770$K given $L = 3.827 \times 10^{33}$ erg/sec and $R = 6.96 \times 10^{10}$ cm.

8.2 Anisotropic Scattering

Derive the equations for the case where the fraction of photons α is scattered forward and $(1 - \alpha)$ is scattered backward.

8.3 Model for Rigel

Rigel is a bright star in Orion. Find its location on the HR diagram in the **See Sky** window. Try building a model that has similar M_{vis} and $B - V$, using a mass of $10 M_{sun}$. Compare your results with the following: $L = 9000$, $R = 65.00$, $distance = 300$ lightyears.

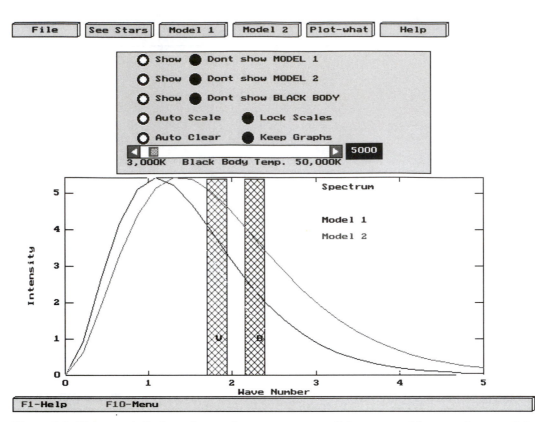

Figure 8.2: This panel displays the continuous spectrum of the two model atmospheres and the spectrum of a blackbody at the temperture specified with the slide bar. Options are selected by clicking on the radio buttons. The two vertical bands show the spectral regions used in defining standard (B-V) colors of stars, and the program evaluates this color for the models and plots the model on the color-magnitude diagram in Figure 8.1.

8.4   Lines of Constant Radius

Using Rigel as the starting point, construct other models with the same M and R but with successively smaller L. In this way, you can trace the lines of constant R on the HR diagram.

Do a similar thing to trace the lines of constant L. You will notice that they are very nearly horizontal. (The luminosity refers to total radiation, and the diagram plots visual magnitude, which does not include all the radiation.)

8.5   Wien Law for Blackbody

With the **Spectrum** window, plot a series of blackbody curves for temperatures 10,000K, 8000K, 6000K, and 4000K. Read off the wave number of maximum intensity. Devise a program (or use a spreadsheet) to plot the wave numbers as a function of temperature and see if you can verify the Wien displacement law, $(1/\lambda)_{max} = cT$.

8.6   Isothermal Atmosphere

a. Compute an isothermal model for a temperature of 6000K. Compare its spectrum with a blackbody at the same temperature.

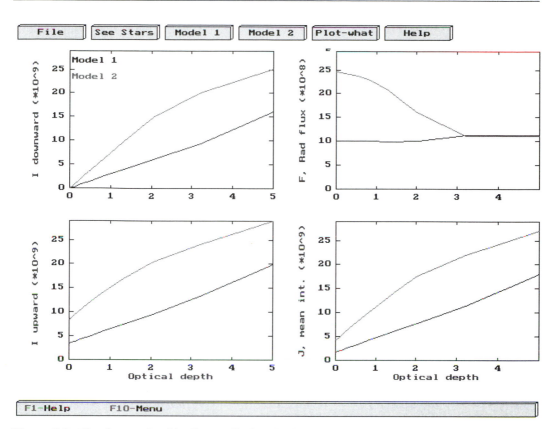

Figure 8.3: The frames in this figure display the internal radiation field as computed from a two-stream approximation in the models. The *x*-coordinate is the (dimensionless) optical depth measured downward from the surface. On the left are shown the upward and downward specific intensities; the upper right shows the net radiative fluxes; the lower right shows the mean intensity, or the average of the upward and downward intensities. In this example, model 2 deviates from radiative equilibrium.

b. Show analytically from the equation of hydrostatic equilibrium that the pressure and density in an isothermal atmosphere increase exponentially with geometric depth. Also, show that the distance over which the pressure and density change by a factor *e* is given by $H = kT/(\mu g)$. Construct models with different *g*, *T*, and helium abundance, and verify this relationship by the slope of logρ plotted against geometric depth.

8.7 Instability Against Convection

a. Use the program to construct a radiative model of the Sun. Notice that the density distribution is slightly flatter than the pressure distribution. Why is this?

b. Now increase the star's luminosity by a factor of 5 and use the program to compute another model. Plot the model and compare its density distribution with the first. You should see that the density actually decreases downward in the lower region of the atmosphere. The density inversion takes place approximately where the radiative temperature gradient exceeds the adiabatic gradient, as you should be able to verify from the graphs. Density inversions are unstable; the gas will turn over, causing convection.

c. Construct similar models with the **IncludeConvection** option and notice that the inversion disappears. What happened to the temperature gradient? Look at the radiation field and see what effect this change of temperature gradient has on the total radiative flux.

8.8 **Variable Flux**

Suppose $H(\tau) = H_0(1 + a \exp(-b\tau))$. Use Eq. 8.45 to find the corresponding source function.

8.9 **Effect of Temperature Distribution**

Build a model for the Sun in radiative equilibrium and verify that the total radiative flux is nearly constant. Now use the **AdjustTemperature** window to distort the temperature distribution and see what this does to the radiative flux. For example, raising the temperature at small optical depths will have very little effect on the flux or the shape of the emitted spectrum. Try intermediate and deep layers. You should be able to correlate your results using the the predictions of Eq. 8.47.

8.6   Suggested Modifications of the Program

1. **Physical Constants**

   Try altering some of the physical constants to see what happens. For example, decrease the ionization potential of hydrogen and see what this does to the density inversion discussed in the exercises.

2. **Variable Gravity**

   Modify the code so you can use a gravity that decreases with depth in the atmosphere. Can you develop an analytical theory that will predict the structure of an isothermal atmosphere with varying gravity? (See ref. 5.)

3. **Starting Procedure**

   Modify the approximation used in Procedure GetStart and show that it will have little effect on the final atmosphere—for reasonable modifications!

8.7   Acknowledgments

Many people have provided comments and assistance along the way. In particular, I wish to thank George Rybicki and Wolfgang Kalkofen for sharing their insights into radiative transfer. Irwin Shapiro, Director of the Harvard-Smithsonian Center for Astrophysics, supported this work, and Philip Sadler, Head of its Science Education Department, provided the environment in which it was performed.

References

1. Auer, L. Improved boundary conditions for the Feutrier method. Astrophysical Journal **150:**L53–L55, 1967.

2. Bohm-Vitense, E. *Introduction to Astrophysics*. Vol. 2, *Stellar Atmospheres* New York: Cambridge University Press, 1989.

3. Collins, G.W. *The Fundamentals of Stellar Astrophysics*. New York: W. H. Freeman, 1989.

4. Kourganoff, V. *Basic Methods in Transfer Problems: Radiative Equilibrium and Neutron Diffusion*. New York: Dover, 1963.

5. Hubeny, I. Influence of radiative transfer on the vertical structure of accretion discs. In *Theory of Accretion Discs*, eds. F. Meyer et al. Boston: Kluwer, 1989.

6. Mihalas, D. *Stellar Atmospheres*. San Francisco: W. H. Freeman, 1970 (2nd ed. 1978).

7. Mihalas, D. The calculation of model stellar atmospheres. In *Methods in Computational Physics*, vol. 7, eds. B. Alder, S. Fernbach, M. Rotenberg. New York: Academic Press, pp. 1–52, 1967.

8. Press, W., Flannery, B. P., Teukolsky, S. A., Vetterling, W.T. *Numerical Recipes: The Art of Scientific Computing*. New York: Cambridge University Press, 1986.

9. Rybicki, G. B., Hummer, D. G. An accelerated lambda iteration method for multilevel radiative transfer. Journal of Astronomy and Astrophysical Journal **245:**171–181, 1991.

10. Rybicki, G. B. A modified Feutrier method. Journal of Quantitative Spectroscopy Radiative Transfer **11:**589–595, 1971.

Appendix A

Walk-Through for All Programs

These "walk-throughs" are intended to give you a quick overview of each program. Please see the Introduction for one-paragraph descriptions for all programs.

A.1   Walk-Through for PLANETS Program

The initial screen is an input data screen, the first of several. This includes data for the total number of planets, and the orbit of Jupiter. Following this, there are separate screens for each planet. During the animation, if you chose to see traces, and these have become too messy, they can be cleaned up by pressing the F5 hot key.

- Start by choosing 10 for the number of planets. Choose to see the traces, but don't make a file to store the animation data. Then select the default values for the planets. Enjoy the animation. When two bodies approach one another closely, the animation will slow down because of the extra calculations that are needed. (Don't begrudge this time: this is where the important physics is taking place!) Note the orbital changes that can result from a close approach, including ejection from the system.

- Hit F10, and return to the sequence of input screens. Change the mass of Jupiter from 0.005 (in units of the mass of the Sun) to 0.05. Keep everything else the same, and repeat the animation, seeing the effect of the higher perturbing mass.

- Rerun, with increased mass for Jupiter. One question regarding planetary systems is whether planets can exist in a binary star system. With Jupiter's mass equal to 0.2, see if you can create a stable system of planets.

- Choose the number of planets to be 2. Restore Jupiter's mass to 0.005, and make its semimajor axis equal to 3 units. For the second planet (which will be interpreted as a comet, here), choose the mass to be zero, the semimajor axis to be

10, and the eccentricity to be 0.7: then the unperturbed perihelion distance will be 3. Choose the starting true anomaly to be 180°, so that the motion will start at aphelion. Run the animation, and see if there is a close encounter with Jupiter. If not, reenter the data, but alter the initial true anomaly of Jupiter. Look for instances where the cometary orbit is transformed to any inside the orbit of Jupiter, or the comet is ejected from the system.

- As another twist on the preceding exercise, choose the semimajor axis of Jupiter's orbit to be 6; for the second planet, choose 3 for its semimajor axis, 0.8 for its eccentricity, and 0 for its starting true anomaly. Rename the Sun to be the Earth, Jupiter to be the Moon, and the second planet to be a spacecraft. By changing the initial true anomaly of the Moon, investigate different sorts of encounters between the Moon and the spacecraft. In particular, look for the "slingshot" effect, by which a close encounter is used to add energy to an orbit.

A.2   Walk-Through ARMS21CM Program

The initial screen shows a default model for a galaxy with two spiral arms already drawn, and the position of the Sun selected. You can redesign the galaxy and can plot more arms, using items in the menu. For now, use this default model.

- Activate the menu by pressing the F10 hot key. Then click on the menu item **Profiles: See 21 cm profiles.** The screen will show three different windows.

- From the figure of the spiral arms, with the ray showing the line of sight, it is clear that parts of four arm segments are observable.

- In the 21 cm line profile, there are four peaks, corresponding to these four arm segments. If you move the cursor into this window, and press the left button, the horizontal coordinates will be shown. Move it to the first peak on the left. The radial velocity is −8.00 km/s.

- Lift the mouse button. This will stop the recording, leaving visible the coordinate of the peak. Move to the window showing the radial velocity profile. Move the cursor, depressed as before, to the point on the curve where the radial velocity is close to −8 km/s. For the value −8.09 km/s, the distance from the Sun is 12.92 kpc.

- Move to the window with the arms. Moving the cursor to the furthest segment, shows that it was that segment that was observed.

- Now repeat with the next peak in the 21 cm profile. There is some ambiguity now about which distance to select from the radial velocity profile. This can be resolved using the illustrated arms. If this illustration was not available, you would have to accumulate data for different galactic longitudes.

- The hot key **Next** will increase the longitude by 15° and **Previous** will reduce it by 15°. **Select** enables you to choose any longitude.

- One option from the menu is to save in a file data for the rotation curve for the galaxy. This is a table of velocities of circular orbits as a function of distance from the galactic center. (After the Pascal listing of this program, you will find a Pascal program for reading this file.) With this information and with a sufficient list of 21 cm profile peaks for different longitudes, it should be possible to reconstruct the arms. See the text for more details.

A.3   Walk-Through ROCHERAD Program

The initial screen is an input data screen. You are prompted to enter the "mass ratio"; that is, the ratio of the mass of one body over the sum of the masses; this is a number between 0 and 1. The default value is 0.5, which means that the masses are equal. You are also prompted to enter parameters for the radiation pressure from each star. There are also quantities relating to the graphics. For the first time, click on **OK** to accept the default values.

- The following screen shows the two stars, at A and B, and five "Lagrangian" equilibrium points. What you are seeing is a reference system rotating with the two stars. You are prompted to enter a value for C. This is the parameter that is used in following different Roche curves. Accept the default value $C = 4$, by clicking on **OK**.

- You will see two curves, the inner one being of the "figure-eight" type. To observe orbits, select the menu item: **Orbit: See an Orbit for This Energy.** Follow the prompt for choosing a starting point and then the direction of the initial velocity. If you try to click for the first time between the curves, you will get a beep, for motion is impossible here. The orbit that you see may look strange; but remember that you are in a rotating reference system. Orbits may pass from one star to the other, but they remain bounded by the figure-eight.

- Use the hot key F10 to return to the menu. Now go back to the data input screen. This time, for the first of the radiation parameters, enter 0.5. Click on **OK**.

- Plot the Roche curves again, for $C = 4$. The figure-eight is not there. One way to look for Roche curves is to use the menu item **New Energy: Click for the Curve Through a Point.** Select this. Then click on a point. Do it again, and see if you can find any figure-eight. Click just above the Lagrangian point between A and B; you will see that a figure-eight is not possible.

- Using this last value of C, start an orbit that moves through the opening from A to B. This orbit is not confined by any Roche curve, and may escape altogether. This would mean that an accretion disk could not be formed.

- We are dealing here with a chaotic dynamical system. This means that small changes in initial conditions may lead to orbits with quite different characteristics. These systems can be studied using Poincaré maps. Before you select that option under the menu item **Orbit**, see the item **Help**, where there are two selections. One provides a short explanation and the other shows a demonstration.

A.4 Walk-Through EVOLVE Program

The program initially comes up with the help screen. A mouse click or keyboard stroke will bring up the menu bar.

- As suggested on the help screen, use the mouse or keyboard controls to **select Getting Started under the File menu item.** This information screen suggests a strategy for learning about the program, which will be followed here. Click the mouse to get back to the menu bar.

- **Select M=1 Star Evolution under the Stages menu item.** This will step you through the stages of evolution for a star of solar mass 1. Each click of the mouse or keyboard will show the next evolution step.

- **Select Show Star Data HR Diagram under the HR Diagram menu item.** The Hertzsprung-Russell (HR) diagram plots stellar luminosity versus surface temperature (or color). The program plot shows the location of computed model stars and actual star observations. Young stars form the zero-age main sequence, with position in the HR diagram only depending on the mass of the star.

- **Select Show HR Diagram Trends under the HR Diagram menu item.** Each click of the mouse or keyboard will show another piece of information about what the HR diagram visualizes about stellar evolution.

- **Select Protostar Evolution under the Protostar menu item.** The plots show the collapse of the protostar proceeding from the inside toward the outside of the initial diffuse cloud.

- **Select Run One Time Step under the Main Sequence menu item.** The program takes a mass 1 star through one convergence process, just as in the STELLAR program. The plots show three of the stellar model parameters plotted versus the integral mass moving out from the center of the star, and an HR diagram that will be used to show the evolution of the star when the model is allowed to change the composition through fusion processes.

- **Select Evolve From ZAMS under the Main Sequence menu item.** The program takes a mass 1 star through a time-sequence of convergence states, with evolution of the star due to nuclear fusion processes.

A.5 Walk-Through STELLAR Program

The program initially comes up with the help screen. A mouse click or keyboard stroke will bring up the menu bar.

- As suggested on the help screen, use the mouse or keyboard controls to **select Getting Started under the File menu item.** This information screen suggests a strategy for learning about the program, which will be followed here. Click the mouse to get back to the menu bar.

- **Select Run Model under the Compute menu item.** This will run a model for a solar mass 1 star. The screen shows plots of four stellar parameters versus the integral mass moving out in the star from the center. An option in the program lets you plot these parameters versus radius rather than mass, which you may find more intuitive. Note how the density and temperature drop slowly moving out in the star from the center, while the luminosity (or integral power generation) rises rapidly. Click the mouse to get back to the menu bar.

- **Select Modify Initial Parameters under the Boundary menu item.** Change the mass value to 1.5 solar masses. Note that this menu lets you change the initial boundary value guesses also. Return to the menu bar, and again select **Run Model** under the **Compute** menu item. The star model is now far from convergence, and you will see the program changing the four boundary conditions until a consistent new equilibrium state is reached. The largest change shows up in the luminosity plot, with smaller changes in the other parameters, showing the extreme temperature dependence of the nuclear reactions in the stellar core. Click the mouse to get back to the menu bar.

- **Again, select Modify Initial Parameters under the Boundary menu item.** Change the mass value to 2 solar masses. Now select **Learn About Boundary Effects** under the **Compute** menu item. This repeatedly runs the computation of the model, but the user must guide the model toward convergence using the hot keys shown at the bottom of the screen. For example, use the **C** hot key to increase the central temperature, the **L** hot key to increase the luminosity, and the **D** hot key to reduce the central density. When you get tired, use **ESC** to exit the computation, and **Select Run Model** under the **Compute** menu item to let the program finish the convergence. With practice, you will find that you will gain an understanding of the boundary parameters and will be able to guide the star close to convergence.

- **Select Plot More Results under the Compute menu item.** Four new plots are shown of other stellar parameters versus mass in the star. Note how the power production is concentrated in the core, but extends at a low level to a large radius. Also note that for this mass 2 star, the CNO cycle is the dominant energy production source. The opacity rises with increasing stellar radius, while the pressure drops. You can see that radiative energy transport dominates in the central part of the star. Clicking on a plot will cause it to be enlarged to fill the screen.

- **Select Plot Power vs Temperature under the Power menu item.** This shows how the nuclear fusion energy rates depend on the temperature. Think about this plot in terms of its effect on properties of stars of different masses. Low mass stars have low core temperatures, while high mass stars have high core temperatures.

- **Select Plot Equation of State under the Eq. State menu item.** These plots, and those seen under **Plot More Equation of State**, show the complex relation of the thermodynamics in the star with temperature and density.

- **Select Plot Opacity under the Opacity menu item.** These plots, and those seen under **Plot More Opacity**, show the complex relation of the opacity of stellar matter with temperature and density.

A.6   Walk-Through ATMOS Program

The program opens with the calculation of two default models and then displays a screen with four panels. These are used for selecting stars whose color and brightness are to be imitated with the models. The lower left shows a menu for selecting a constellation or the nearby stars. Select Orion. The upper panel shows a map of the brightest stars in Orion. Click on one of them to select it. You will see a circle drawn about the selected star. In the upper right panel, the same star will be circled. Try clicking in either of the upper panels and guess which image will be circled in the other window. Each time you select a star, its data are displayed in the lower right panel. For example, when you click on the lower right star in Orion, you will see the name Rigel in the lower right panel along with other data for this star.

The upper right panel is called a color-magnitude diagram because each star is plotted according to its photometric color and its absolute magnitude. These parameters are measures of temperature and brightness. Hot, blue stars are to the left; brighter stars are upward.

One of the challenges offered by this program is to build a model atmosphere that has properties similar to those of the star you have selected.

Find a star you would like to imitate, and notice where it lies in the color-magnitude diagram. As an example, shift to the constellation Scorpio and select the third star from the end of the tail, eta Scorpii. You will see that the default model of Vega is very close to this star, while the model for the Sun is lower (fainter) and slightly to the right (hotter).

Now select the star just to the left of eta. It is Sargas and it is brighter than the Sun, but has nearly the same color. Let us take Sargas as the star to be imitated.

Building a Gray Model Atmosphere:

1. Select the **Model 1/Specify Model** item.

2. Enter the name, Sargas, and the following guesses for the mass radius and luminosity: $M = 9$, $R = 10$, $L = 100$. (Leave the distance where it is, for the moment.) For the type of structure, select **Pure Radiative**. This excludes convection and computes the initial temperatures from the theoretical distribution for a gray atmosphere. Finally, select **Gray** opacity.

3. Close the window by clicking on the **OK** button or hitting **Return**. The model is automatically built and is now ready for plotting.

Examining the Location of the Model on the Color-Magnitude Diagram

1. For the moment we wish to see only the data plotted for a single model, so select the **Plot What/Plot Options** item and deselect the button for plotting model 2 only.

2. Select the **See Stars/On Sky** item and notice where model 1 is plotted on the color-magnitude diagram. In this case, the model lies below the real data for Sargas.

3. To improve the agreement with the data, increase the luminosity to 520 suns and the radius to 16 suns. Experiment with different values and see how the star moves on the color-magnitude diagram.

Comparing the Model Spectrum With a Black Body Spectrum

When the gray model gives a satisfactory fit on the color-magnitude diagram, you may examine its spectrum and compare it with a black body spectrum. (It is not possible to call up the spectrum of the real star.)

Examining the Structure of the Model

The effects of convection: A density inversion will cause the atmosphere to be unstable against convection, and the calculation of temperature structure ought to include the effect of convective transport of energy from the deeper layers toward the surface. The program can include a mixing-length theory for convection.

1. Select the **Model 2/Specify Model** and enter the name Sargas C.

2. Enter the same parameters as for model 1, except for the **Type of Structure,** where you should select **Incl. Convection.**

3. Hit the **OK** button and select the **See Stars/Spectrograph.** You will see that the spectrum of the new model is nearly identical to that of the purely radiative model.

4. Select **Plot What/Gas Structure** and compare the two models. You will see that the density inversion is less severe and the temperature is slightly lower at great depths in model 2, due to the convection.

5. Select **Plot What/Radiation Field** and you will see that the radiative flux in model 2 decreases at optical depths greater than unity. The difference is carried by convection.

6. Explore the shapes of the various plots of gas structure. For example, with the **Plot What/Plot Options,** change the x-coordinate to geometric depth and notice that the **Log(P)** plot is nearly linear because the temperature changes relatively little in the upper layers (about 0.2 in the logarithm). Also, verify that the density inversion occurs where the opacity increases, and this is also where hydrogen is becoming ionized.

Examining the Effects of Hydrogen Opacity

We want to compare two models for Sargas that are identical except for the opacity calculation. This will reveal the effect of the wavelength variation of the opacity of hydrogen, which is the most common element in most stellar atmospheres.

1. Select **Model 1/Specify Model** and enter the values of *M*, *R*, and *L*; select **Gray** opacity and **Incl. Convection.**

2. Select **Model 2/Specify Model** and enter the values of *M*, *R*, and *L*; select **Hydr.** opacity and **Incl. Convection.**

3. Select **See Stars/On Sky** and notice that they have the same brightness, but the model with **Hydr.** opacity is slightly to the left, or bluer than the gray model.

4. To see why this is the case, select **See Stars/Spectrograph** and look at the spectra of the two models. Notice that in the region of the B and V filters, used for defining the *x*-coordinate in the color-magnitude diagram, the gray model slopes down more steeply to the blue than the model with hydrogen opacity. This gives the gray model a slightly redder color and moves it to the left in the color-magnitude diagram.

5. With the **Lock Scales** option, compare the two spectra with the spectrum of a black body at various temperatures.

A word about the star data. The constellation maps and star data are stored in a separate file called STARDATA. It is in record format and cannot be directly modified. A separate program is provided, called CREMAIN, that permits creating a file in the proper REC format.

A.7   Walk-Through for PULSE Program

The program PULSE, written by Charles Whitney, can be used for exploring the basic structure of stellar models and some of their pulsation properties. Several types of interior models can be constructed. The program solves for the radial modes of adiabatic oscillation. All of the pulsation calculations are based on the linearized equations for adiabatic motion. This means that the program is limited to displaying acoustic wave functions and determining the modal periods; it cannot be used for a study of damping and instability.

When the program opens, a default model for the Sun is constructed and a varied set of graphs is opened. The pulsation in the fundamental mode is displayed. You may change the pace of the simulation by using the function keys, as indicated at the bottom of the screen. Try making it go faster. Each time you hit the faster function key, the time step is doubled. The pulsation is computed by Fourier synthesis, so there is no need to worry about taking too large a step size.

Changing the Graphical Display

The default shows four types of graphs: animation of the cross section for a selected variable, strip chart, profile graph, and phase diagram in which two variables are plotted against each other. With the **Plot What/Type of graph** item, you can select any set of four graphs you wish; with the **/Variable** item, you can select which of the thermodynamic variables will be plotted on each graph.

As an example, put two line graphs on the left and two strip charts on the right. Let the left graphs display displacement and the others display velocity. Restart the simulation with the function key or the **Run/Start** item and see if you can understand the relations between the four graphs.

Pulsation in Mixed Modes

So far, you have been seeing the fundamental mode, in which the center is a node and the surface is an antinode. Now, with the **Run/Start** from **Modes** item, open the window that permits establishing arbitrary phases and amplitudes for the lowest four modes. By clicking the radio buttons and entering numerical data in the appropriate places you can establish any desired combination of these modes.

As an example, click the radio buttons so the fundamental is inactive and the first overtone is active. (You may also set the initial phase of each mode if you wish, but these phases constantly change as the periods are incommensurate with each other in a typical model. One of the advanced exercises is to see if you can build a model in which the fundamental and second overtone have a 2:1 period ratio.)

With the **Model/See Modes** item you can display the active modes and their periods and amplitudes.

Start the pulsation and notice that the interior node is very close to the center of the star. This is a result of the concentration of density toward the center. Before building a different model, we will look at the structure of this model.

Examining the Underlying Stellar Model

With the **Model/Plot Structure** and **Plot More Structure** items you can see plots of the underlying structure of the static star. These models are incomplete, in that they are models of the envelope only; they do not go all the way to the center of the star. For our purpose that is quite adequate and it avoids the need for a search procedure, as is used in the STELLAR INTERIOR program by Richard Kouzes.

Building a Homogeneous Model

If you select the **Model/Specify Build** item, you can select new stellar parameters and explore the changes of the periods and shapes of the pulsation modes. One useful type is the homogenous—constant density—model, which you can build by selecting the appropriate radio button in the **Specify/Build** window and closing the window by selecting the **Model/Plot** structure item. When the graphs are displayed, you will see that the density is constant, while the pressure (which is computed from the differential equation of hydrostatic equilibrium) is a quadratic function of distance. You might want to try verifying the shape of the plotted curve by deriving an expression for the pressure distribution from the equation of hydrostatic equilibrium, $\frac{dP}{dr} = -GM(r)\rho/r^2$. What can you say about the temperature distribution?

Wavefunctions for the Homogeneous Model

Construct profile plots for the displacement, pressure, and velocity. (If you make sure there are no strip charts or cross-sectional animations, the simulation will go faster.)

The fundamental mode has a particularly simple shape in the homogeneous model: namely, it is linear. Make sure you see why the three plots are related to each other the way they are. Now select the **Run/Start** from the **Modes** item and examine the successively higher modes. Note they are easier to see than for the model of the Sun, due to a more favorable density distribution, which enhances the amplitudes in the deep interior.

Appendix B

Mechanics

J. M. Anthony Danby

The material contained in this appendix is designed as a summary of the mechanics background and the methods of computation used in the programs on binaries and *n*-bodies. If you want to modify the programs, or write similar ones, you may need this material.

B.1 Newton's Law of Gravitation

Let there be two particles (or "point masses"), m_1 at the point P_1 and m_2 at P_2. For any origin O, let $\overrightarrow{OP}_1 = \vec{r}_1$, and $\overrightarrow{OP}_2 = \vec{r}_2$. Newton's law of gravitation states that the gravitational force exerted by m_2 on m_1 is proportional to the product of the masses, inversely proportional to the square of the distance between them, and directed in the line joining the masses. Then this force can be written as

$$\vec{F} = Gm_1m_2 \frac{1}{r_{12}^2} \frac{\vec{r}_2 - \vec{r}_1}{r_{12}} = Gm_1m_2 \frac{\vec{r}_2 - \vec{r}_1}{r_{12}^3}. \tag{B.1}$$

Here $r_{12} = \|\vec{r}_2 - \vec{r}_1\|$, the distance between the bodies. The value of the constant of gravitation, G, depends on the units used.

The words "particle" or "point masses" are convenient for setting up equations. In many instances, astronomical bodies can be assumed to be spherically symmetrical. In this case, for gravitational purposes, they can be treated as particles at points exterior to the bodies.

B.2 Equations of Motion of N-Bodies

Let O be the origin of an inertial system of reference. Consider a system of particles, m_i at P_i, where $\overrightarrow{OP}_i = \vec{r}_i$ and $i = 1 \ldots n$. From Newton's second law, the equation of motion of m_i is

$$m_i \frac{d^2\vec{r}_i}{dt^2} = G \sum_{\substack{j=1 \\ j\neq i}}^{n} m_i m_j \frac{\vec{r}_j - \vec{r}_i}{r_{ij}^3}, \tag{B.2}$$

or

$$\frac{d^2\vec{r}_i}{dt^2} = G \sum_{\substack{j=1 \\ j\neq i}}^{n} m_j \frac{\vec{r}_j - \vec{r}_i}{r_{ij}^3}, \quad i = 1 \dots n. \tag{B.3}$$

If all of the n equations (Eq. B.2) are added, the terms on the right vanish. The terms on the left can be integrated twice, to give

$$\sum_{i=1}^{n} m_i \vec{r}_i = \vec{A}t + \vec{B}, \tag{B.4}$$

where $\vec{A}$ and $\vec{B}$ are arbitrary constants. Let the center of mass of the system be at $\vec{r}_g$. Then

$$\vec{r}_g = \frac{\left(\sum_{i=1}^{n} m_i \vec{r}_i \right)}{\sum_{i=1}^{n} m_i}. \tag{B.5}$$

So it has been established that the center of mass of the system is not accelerated with respect to the chosen inertial reference system. Consequently, the center of mass is itself a valid origin of an inertial reference system.

It is often convenient to transfer the origin to one of the particles. For instance, in planetary work it is usual to have the origin at the center of the Sun. Let us transfer the origin to the mass m_n. Let the vector from m_n to m_j be

$$\vec{R}_j = \vec{r}_j - \vec{r}_n. \tag{B.6}$$

Then the equation of motion of m_i becomes

$$\begin{aligned}
\frac{d^2\vec{R}_i}{dt^2} &= \frac{d^2\vec{r}_i}{dt^2} - \frac{d^2\vec{r}_n}{dt^2} \\
&= G \sum_{\substack{j=1 \\ j\neq i}}^{n} m_j \frac{\vec{r}_j - \vec{r}_i}{r_{ij}^3} - G \sum_{j=1}^{n-1} m_j \frac{\vec{r}_j - \vec{r}_n}{r_{jn}^3} \\
&= G \sum_{\substack{j=1 \\ j\neq i}}^{n-1} m_j \left(\frac{\vec{R}_j - \vec{R}_i}{R_{ij}^3} - \frac{\vec{R}_j}{R_j^3} \right) - G(m_n + m_i) \frac{\vec{R}_i}{R_i^3}.
\end{aligned} \tag{B.7}$$

This equation can be written as

$$\frac{d^2\vec{R}_i}{dt^2} + G(m_n + m_i) \frac{\vec{R}_i}{R_i^3} = G \sum_{\substack{j=1 \\ j\neq i}}^{n-1} m_j \left(\frac{\vec{R}_j - \vec{R}_i}{R_{ij}^3} - \frac{\vec{R}_j}{R_j^3} \right). \tag{B.8}$$

The components of the equation can be discussed as follows. If all masses but m_i and m_n are zero, the right-hand side is zero, and there remain the equations of motion for two bodies. More generally, the terms on the right can be interpreted as forces perturbing the motion of two bodies, a situation often arising in dynamical astronomy. Of the two terms appearing in parentheses on the right, the first is the

direct attraction of m_j on m_i. The second is called **indirect**, and is the attraction of m_j on m_n at the *non-inertial* origin.

B.3 Equations of Motion for Two Bodies

Let the masses be m_1 and m_2 and let the vector from m_1 to m_2 be $\vec{r}$. Then

$$\frac{d^2\vec{r}}{dt^2} = -G(m_1 + m_2)\frac{\vec{r}}{r^3}, \tag{B.9}$$

where $r = \|\vec{r}\|$. This can be written as

$$\frac{m_1 m_2}{m_1 + m_2}\frac{d^2\vec{r}}{dt^2} = -Gm_1 m_2 \frac{\vec{r}}{r^3}. \tag{B.10}$$

On the right, we have the mutual gravitational force. Since the origin is non-inertial, Newton's second law is not applicable; but its applicability can be *pretended* if the "reduced mass," $m_1 m_2/(m_1 + m_2)$, replaces the true mass on the **left** side of the equation. Certain derivations become more straightforward if the reduced mass is used.

In some two-body systems, such as spectroscopic binaries, it is necessary to have the origin at the center of mass of the system. Let O be at the center of mass, with position vectors to the masses $\vec{r}_1$ and $\vec{r}_2$. Since

$$\vec{r}_2 = \frac{m_1}{m_1 + m_2}\vec{r}, \tag{B.11}$$

the equation of motion of m_2 is

$$\frac{d^2\vec{r}_2}{dt^2} = -G\frac{m_1^3}{(m_1 + m_2)^2}\frac{\vec{r}_2}{r_2^3}, \tag{B.12}$$

where Eq. B.11 was substituted into Eq. B.9. Similarly, for m_1,

$$\frac{d^2\vec{r}_1}{dt^2} = -G\frac{m_2^3}{(m_1 + m_2)^2}\frac{\vec{r}_1}{r_1^3}. \tag{B.13}$$

All of these equations are of the form

$$\frac{d^2\vec{r}}{dt^2} = -\mu\frac{\vec{r}}{r^3}, \tag{B.14}$$

which will be the notation used below.

B.4 Keplerian Motion

When "Kepler's laws" are applied in astronomy, it is in their post-Newtonian form. That is, the first law is expanded to allow parabolic and hyperbolic orbits and even degenerate cases like the "linear ellipse," which is not among the mathematical family of conics. The third law is modified to relate the period of a revolution to the size of the orbit **and** the masses. If the period is P, then

$$P = 2\pi \sqrt{\frac{a^3}{G(m_2 + m_2)}}. \tag{B.15}$$

Here a is the *semimajor axis* of the ellipse.

If e is the *eccentricity* of the ellipse, then $0 \le e < 1$, where $e = 0$ corresponds to a circle. The *semiminor axis*, $b = a\sqrt{1 - e^2}$. The end of the major axis closest to the attracting focus is the "pericenter," and that furthest from the attracting focus is the "apocenter." The *pericenter distance* is $q = a(1 - e)$ and the *apocenter distance* is $q' = a(1 + e)$. The word "center" is all-purpose. For motion around the Sun, one can use "perihelion"; around the Earth, "perigee"; around a star, "periastron"; and around a galaxy, "perigalacticon." From Eq. B.14, we have

$$\vec{r} \times \frac{d^2\vec{r}}{dt^2} = -\mu\vec{r} \times \frac{\vec{r}}{r^3} = \vec{0}, \tag{B.16}$$

since $\vec{r} \times \vec{r} = \vec{0}$. Therefore,

$$\vec{r} \times \frac{d\vec{r}}{dt} = \vec{r} \times \vec{v} = \vec{h}, \tag{B.17}$$

a constant. This is equivalent to the conservation of orbital angular momentum. Provided that $\vec{h} \ne \vec{0}$, the orbit lies in a plane perpendicular to $\vec{h}$. (If $\vec{h} = \vec{0}$, then the orbit is linear, and not of astronomical importance. This possibility will be ignored here.)

Again, from Eq. B.14, and using $\vec{v} = d\vec{r}/dt$,

$$\vec{v} \cdot \frac{d\vec{v}}{dt} = -\mu\vec{v} \cdot \frac{\vec{r}}{r^3} = \mu\frac{d}{dt}\left(\frac{1}{r}\right). \tag{B.18}$$

So

$$\frac{1}{2}v^2 = \frac{\mu}{r} + constant. \tag{B.19}$$

This is equivalent to the conservation of energy.

Let $\hat{r}$ be the unit vector parallel to $\vec{r}$. Then

$$\frac{d\hat{r}}{dt} = \frac{d}{dt}\left(\frac{\vec{r}}{r}\right)$$

$$= \frac{1}{r^2}\left(\frac{d\vec{r}}{dt}r - \vec{r}\frac{dr}{dt}\right)$$

$$= \frac{1}{r^3}\left(\frac{d\vec{r}}{dt}\vec{r}\cdot\vec{r} - \vec{r}\left(\vec{r}\cdot\frac{d\vec{r}}{dt}\right)\right)$$

$$= \frac{1}{r^3}(\vec{r} \times \vec{v}) \times \vec{r}$$

$$= \frac{\vec{h} \times \vec{r}}{r^3}. \tag{B.20}$$

Here we have used the formula for the triple vector product,

$$(\vec{a} \times \vec{b}) \times \vec{c} = (\vec{c} \cdot \vec{a})\vec{b} - (\vec{c} \cdot \vec{b})\vec{a},$$ (B.21)

and the useful formula

$$\vec{r} \cdot \frac{d\vec{r}}{dt} = r\frac{dr}{dt}.$$ (B.22)

It follows, again from Eq. B.14, that

$$\frac{d^2\vec{r}}{dt^2} \times \vec{h} = \mu\frac{d\hat{r}}{dt},$$ (B.23)

so, integrating,

$$\vec{v} \times \vec{h} = \mu(\hat{r} + \vec{e}).$$ (B.24)

The constant vector $\vec{e}$ lies in the plane of the orbit. Next, multiply each side by $\vec{r}$. Now

$$\vec{r} \cdot (\vec{v} \times \vec{h}) = (\vec{r} \times \vec{v}) \cdot \vec{h} = h^2.$$ (B.25)

So

$$h^2 = \mu(r + \vec{e} \cdot \vec{r}),$$ (B.26)

or

$$\frac{h^2/\mu}{r} = 1 + e\cos f.$$ (B.27)

f is the angle measured from the direction of $\vec{e}$ to that of $\vec{r}$; it is called the *true anomaly*, and is shown in Figure B.1. Equation B.27 is the polar equation of a conic; the direction of $\vec{e}$ is the direction of pericenter.

$$p = \frac{h^2}{\mu} = a(1 - e^2)$$ (B.28)

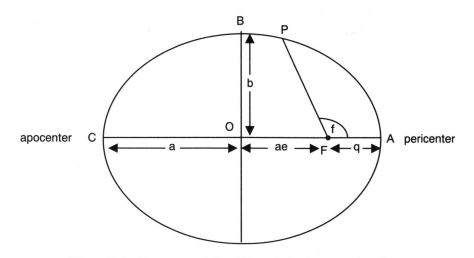

Figure B.1: Geometry of the ellipse. F is the attracting focus.

is called the *parameter* or *semilatus rectum* of the conic. If we square both sides of Eq. B.24 and use Eq. B.27, we find

$$v^2 = \mu \left(\frac{2}{r} - \frac{1}{a} \right), \qquad (B.29)$$

to be compared with Eq. B.19.

The angular momentum, per unit mass, is $r^2 \, df/dt$, so we have the important relation

$$h = r^2 \frac{df}{dt}. \qquad (B.30)$$

By elementary calculus, the area swept out by the radius vector from time t_0 to time t is

$$A(t) = \frac{1}{2} \int_{t_0}^{t} r^2 \, df, \qquad (B.31)$$

so

$$\frac{dA}{dt} = \frac{1}{2} h. \qquad (B.32)$$

These equations form the solution of Eq. B.14 and are valid for elliptic, parabolic, or hyperbolic orbits. For hyperbolic orbits it is usual to take a to be negative.

B.5   Elliptic Motion

The discussion that follows is valid only for bound, elliptic orbits.

The area of an ellipse is $\pi ab = \pi a^2 \sqrt{1 - e^2}$. So, from Eq. B.32, if P is the time taken for a complete revolution,

$$\pi a^2 \sqrt{1 - e^2} = \frac{1}{2} hP,$$

consistent with Eq. B.15.

To introduce the time into the solution it is necessary to define an intermediate angle, the *eccentric anomaly*, E. This is defined, geometrically, in Figure B.2. An ellipse can be defined as the projection of a circle, the *auxiliary circle*. This is drawn in the figure along with the ellipse. A point P on the ellipse corresponds to a point R on the circle; the line RP is perpendicular to the major axis and intersects the major axis at the point Q. F is the attracting focus, A is at pericenter, and O is at the center of the ellipse. The angle $\angle AFP = f$, the true anomaly; and $\angle AOR = E$, the eccentric anomaly. See Figure B.2.

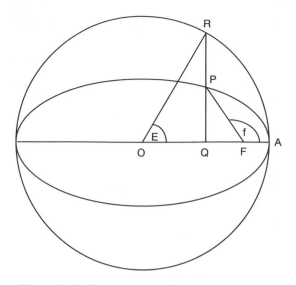

Figure B.2: The eccentric and true anomalies.

We shall apply the law of areas:

$$\frac{\text{time in the orbit from } A \text{ to } P}{\text{time of a complete revolution}} = \frac{\text{area of sector } AFP}{\text{area of ellipse}}$$

$$= \frac{\text{area of sector } AFR}{\text{area of circle}}$$

$$= \frac{\text{area of Sector } AOR - \text{area of triangle } FOR}{2\pi a^2}$$

$$= \frac{1}{2\pi a^2}\left(\frac{1}{2}a^2 E - \frac{1}{2}ae(a\sin E)\right)$$

$$= \frac{1}{2\pi}(E - e\sin E). \tag{B.33}$$

Define the frequency

$$n = \frac{2\pi}{P} = \sqrt{\frac{\mu}{a^3}}, \tag{B.34}$$

called the *mean motion*. Let the orbiting body be at A at time T, called the *time of pericenter*, and at P at time t. Then

$$M = n(t - T) \tag{B.35}$$

is called the *mean anomaly*. The time from A to P divided by the period of the orbit is, then, $M/2\pi$. So we have

$$M = n(t - T) = E - e\sin E, \tag{B.36}$$

which is *Kepler's equation*.

As an intermediate stage in many calculations it is convenient to find coordinates in the *orbital reference system*. See Figure B.3. This has the origin at the

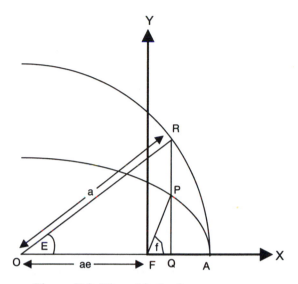

Figure B.3: The orbital reference system.

attracting focus, and the X-axis pointing toward pericenter, as shown. In terms of the true or eccentric anomalies, the coordinates and their time derivatives can be expressed as

$$
\left.
\begin{aligned}
X &= r \cos f & &= a(\cos E - e) \\
Y &= r \sin f & &= a\sqrt{1 - e^2}\, \sin E \\
\dot{X} &= -\frac{\mu}{h} \sin f & &= -\frac{na^2}{r} \sin E \\
\dot{Y} &= \frac{\mu}{h}(e + \cos f) &= \frac{na^2 \sqrt{1 - e^2}}{r} \cos E \\
r & & &= a(1 - e \cos E) \\
\dot{r} &= \frac{e\mu}{h} \sin f & &= \frac{ena^2}{r} \sin E
\end{aligned}
\right].
\tag{B.37}
$$

Many computations relate position and velocity at different times. If $\vec{r}_0$ and $\vec{v}_0$ are position and velocity vectors at time t_0, and $\vec{r}$ and $\vec{v}$ are similar vectors at time t, then, because the orbit takes place in a plane, there exist unique scalars, **f** and **g**, such that

$$
\vec{r} = \mathbf{f}\vec{r}_0 + \mathbf{g}\vec{v}_0.
\tag{B.38}
$$

The **f** and **g** functions play a central role in computations involving Keplerian motion. They are functions of t and t_0. If

$$
\dot{\mathbf{f}} = \frac{\partial \mathbf{f}}{\partial t} \quad \text{and} \quad \dot{\mathbf{g}} = \frac{\partial \mathbf{g}}{\partial t},
\tag{B.39}
$$

then

$$
\frac{d\vec{r}}{dt} = \dot{\mathbf{f}}\vec{r}_0 + \dot{\mathbf{g}}\vec{v}_0.
\tag{B.40}
$$

An advantage of the use of these functions is that these formulas are independent of the reference system that is used. Formulas for the functions can be derived by resolving Eqs. B.38 and B.40 in the orbital reference system and solving for the functions and their derivatives. The results are

$$\mathbf{f} = \frac{a}{r_0}[\cos(E - E_0) - 1] + 1,$$

$$\mathbf{g} = \frac{1}{n}[\sin(E - E_0) - e \sin E + e \sin E_0]$$

$$= t - t_0 + \frac{1}{n}[\sin(E - E_0) - (E - E_0)].$$

$$\dot{\mathbf{f}} = -\frac{a}{r}\frac{a}{r_0}n \sin(E - E_0),$$

$$\dot{\mathbf{g}} = \frac{a}{r}[\cos(E - E_0) - 1] + 1.$$

$$\tag{B.41}$$

$$\mathbf{f} = \frac{r}{p}[\cos(f - f_0) - 1] + 1,$$

$$\mathbf{g} = \frac{rr_0}{\sqrt{\mu p}}\sin(f - f_0),$$

$$\dot{\mathbf{f}} = -\sqrt{\frac{\mu}{p^3}}[\sin(f - f_0) + e \sin f - e \sin f_0],$$

$$\dot{\mathbf{g}} = \frac{r_0}{p}[\cos(f - f_0) - 1] + 1.$$

$$\tag{B.42}$$

The **f** and **g** *functions* only came into prominence in the 1950s with the widespread use of high-speed computation. Up till then, they were **f** and **g** *series*, expanded in powers of the time difference $(t - t_0)$. We have the Taylor series:

$$\vec{r}(t) = \vec{r}_0 + (t - t_0)\dot{\vec{r}}_0 + \frac{1}{2}(t - t_0)^2\ddot{\vec{r}}_0 + \frac{1}{6}(t - t_0)^3\dddot{\vec{r}}_0 + \cdots \tag{B.43}$$

Now

$$\ddot{\vec{r}}_0 = -\frac{\mu\vec{r}_0}{r_0^3} = -\sigma\vec{r}_0,$$

and

$$\dddot{\vec{r}}_0 = -\frac{\mu\vec{v}_0}{r_0^3} + \frac{3\mu\vec{r}_0\dot{r}_0}{r_0^4} = -\sigma\vec{v}_0 + 3\sigma\tau\vec{r}_0,$$

where

$$\sigma = \frac{\mu}{r_0^3} \quad \text{and} \quad \tau = \frac{\dot{r}_0}{r_0} = \frac{\vec{r}_0 \cdot \vec{v}_0}{r_0^2}.$$

Substituting, and rearranging, we find

$$\vec{r}(t) = \left(1 - \frac{1}{2}\sigma(t - t_0)^2 + \frac{1}{2}\sigma\tau(t - t_0)^3 + \cdots\right)\vec{r}_0$$
$$+ \left((t - t_0) - \frac{1}{6}\sigma(t - t_0)^3 + \cdots\right)\vec{v}_0.$$

Comparing this with Eq. B.38, we can write

$$\left.\begin{array}{l} \mathbf{f} = 1 - \dfrac{1}{2}\sigma(t - t_0)^2 + \dfrac{1}{2}\sigma\tau(t - t_0)^3 + \cdots, \\[1em] \mathbf{g} = (t - t_0) - \dfrac{1}{6}\sigma(t - t_0)^3 + \cdots. \end{array}\right] \tag{B.44}$$

These are the **f** and **g** *series*. They can be very useful if computations over short time intervals are required, and are valid for any type of orbit. They can be developed to higher powers of $(t - t_0)$, but from a computational point of view, if series are required, they should be developed recursively, as will be described later. These series are used in the program for colliding galaxies.

B.6   Universal Variables

Sometimes one set of formulas is required that is valid for any type of orbit. (This is the case in the program for the colliding galaxies.) Then *universal variables* should be used. A universal variable or a universal equation is one that remains valid for elliptic, parabolic, or hyperbolic orbits. A change of independent variable is made to the new variable s. s is set equal to zero for the time $t = t_0$. So that

$$dt = r\,ds \tag{B.45}$$

and

$$t - t_0 = \int_0^s r(\bar{s})\,d\bar{s}. \tag{B.46}$$

In terms of this new variable the singularity at $r = 0$ does not occur. The radius satisfies the differential equation

$$\frac{d^2r}{ds^2} + \alpha r = \mu, \quad \text{where} \quad \alpha = \frac{\mu}{a} = \frac{2\mu}{r} - \vec{v}^2, \tag{B.47}$$

and the **f** and **g** functions satisfy similar equations. These equations might be solved for the separate cases $\alpha > 0, < 0, = 0$. But this is precisely what is to be avoided if universality is required. A set of functions is introduced called **Stumpff functions**, after their originator; these satisfy the equations and can be calculated without regard to the sign of α. A new form of Kepler's equation can then be derived, using these functions. Full details, together with programming suggestions, will be found in J. M. A. Danby,[1] sections 6.9 and 6.10.

B.7   The Solution of Kepler's Equation

We next turn to practical concerns of computation. Kepler's equation, Eq. B.36, or a modification of it, may be solved many thousands of times during the execution of a program, so its solution should be coded with some care. It can be written as

$$f(x) = x - e \sin x - M = 0 . \tag{B.48}$$

$f'(x) > 0$; so, for given e and M, there is just one solution for x. This will be approached by a sequence of successive approximations.

Let us call the actual solution $x = a$, and the sequence of approximations, $\{x_n\}$, $n = 0, 1, 2, \ldots$. Let the error of the approximation x_n be ϵ_n, so that

$$x_n + \epsilon_n = a . \tag{B.49}$$

Then we can generate the Taylor series expansion:

$$0 = f(a) = f(x_n + \epsilon_n)$$

$$= f(x_n) + \epsilon_n f'(x_n) + \frac{1}{2} \epsilon_n^2 f''(x_n) + \frac{1}{6} \epsilon_n^3 f'''(x_n) + \cdots \tag{B.50}$$

Let us define δ_n by

$$0 = f(x_n) + \delta_n f'(x_n) + \frac{1}{2} \delta_n^2 f''(x_n) + \frac{1}{6} \delta_n^3 f'''(x_n) ,$$

which will be abbreviated as

$$0 = f_n + \delta_n f'_n + \frac{1}{2} \delta_n^2 f''_n + \frac{1}{6} \delta_n^3 f'''_n . \tag{B.51}$$

This can be written in the form

$$\delta_n = - \frac{f_n}{f'_n + \frac{1}{2} \delta_n f''_n + \frac{1}{6} \delta_n^3 f'''_n} , \tag{B.52}$$

and this can be solved for δ_n with an accuracy consistent with the approximation made in truncating the series in Eq. B.50 by:

$$\delta_{n1} = - \frac{f_n}{f'_n} ,$$

$$\delta_{n2} = - \frac{f_n}{f'_n + \frac{1}{2} \delta_{n1} f''_n} ,$$

$$\delta_{n3} = - \frac{f_n}{f'_n + \frac{1}{2} \delta_{n2} f''_n + \frac{1}{6} \delta_{n2}^2 f'''_n} \tag{B.53}$$

followed by

$$x_{n+1} = x_n + \delta_{n3} ,$$

which leads to a sequence $\{x_n\}$ having (in theory) quartic convergence. (That is, the error of x_{n+1} depends on the *fourth* power of the error of x_n.) If just δ_{n1} were used, we should have Newton's method, with quadratic convergence. Stopping at δ_{n2} would give Halley's method, with cubic convergence. For an equation as benign as Kepler's equation, this procedure works well, and is considerably faster than Newton's method.

It is important to start with a good initial approximation, x_0. For nearly all purposes, I find it efficient to use

$$x_0 = M + ke, \quad 0 \le M \le \pi, \quad \text{and} \quad x_0 = M - ke, \quad -\pi \le M < 0. \quad \text{(B.54)}$$

$k = 0.8$ is satisfactory. Before solving Kepler's equation M should be increased or decreased by a multiple of 2π so that it lies within the interval needed for the application of Eq. B.54.

B.8   The Initial Value Problem

The problem to be addressed is "Given initial conditions $\vec{r}_0 = \vec{r}(t_0)$ and $\vec{v}_0 = \vec{v}(t_0)$ for a time t_0, and given a second time t, find $\vec{r}(t)$ and $\vec{v}(t)$." We shall consider only those initial conditions that lead to an elliptic orbit.

From the initial conditions we can calculate $r_0 = |\vec{r}_0|$, v_0^2, and the scalar product

$$u = \vec{r}_0 \cdot \vec{v}_0 = r_0 \dot{r}_0. \quad \text{(B.55)}$$

From the energy integral, Eq. B.29, we can solve for a:

$$a = \left(\frac{2}{r_0} - \frac{v_0^2}{\mu} \right)^{-1}. \quad \text{(B.56)}$$

If a is negative, the initial conditions are invalid, since they would result in a hyperbolic orbit. Otherwise, the mean motion n is

$$n = \sqrt{\frac{\mu}{a^3}}. \quad \text{(B.57)}$$

Next, from Eq. B.37 we can show that

$$r = a(1 - e \cos E), \quad \text{(B.58)}$$

and

$$r \frac{dr}{dt} = ena^2 \sin E. \quad \text{(B.59)}$$

Then at the initial time t_0

$$r_0 = a(1 - e \cos E_0) \quad \text{and} \quad u = r_0 \dot{r}_0 = ena^2 \sin E_0.$$

It is convenient to define two new variables EC and ES, where

$$EC \equiv e \cos E_0 = 1 - r_0/a,$$

and

$$ES \equiv e \sin E_0 = u/na^2. \quad \text{(B.60)}$$

Note that e can be found from these; it will be needed when making an initial guess for the solution of Kepler's equation.

Writing down Kepler's equation for the times t_0 and t, and subtracting one from the other, see Eq. B.36, we find

$$n\Delta t = \Delta E - e \sin E + e \sin E_0, \tag{B.61}$$

where

$$\Delta t = t - t_0, \quad \text{and} \quad \Delta E = E - E_0. \tag{B.62}$$

Then $E = \Delta E + E_0$, so Eq. B.61 can be rewritten as

$$n\Delta t = \Delta E - e \cos E_0 \sin \Delta E + e \sin E_0(1 - \cos \Delta E). \tag{B.63}$$

This is the new form for Kepler's equation. Notice that for a circular orbit the difference ΔE can always be found in spite of the fact that individual anomalies are not defined. Eq. B.63 is to be solved for the unknown $x = \Delta E$. The solution of

$$f(x) = x - EC \sin x + ES(1 - \cos x) - n\Delta t = 0 \tag{B.64}$$

is as straightforward as that for the basic form of Kepler's equation. But the initial guess x_0 requires more care. Since

$$x = \Delta E = E - E_0,$$

then

$$x_0 = (M + \sigma ke) - (M_0 + e \sin E_0)$$
$$= n\Delta t + \sigma ke - e \sin E_0. \tag{B.65}$$

To identify σ we need to know the sign of $\sin M$. Now

$$M = n(t - T)$$
$$= n(t - t_0) + n(t_0 - T)$$
$$= n\Delta t + (E_0 - e \sin E_0).$$

We want to avoid explicitly finding E_0. But we can write

$$\sin M = \sin E_0 \cos(n\Delta t - e \sin E_0) + \cos E_0 \sin(n\Delta t - e \sin E_0)$$
$$= \frac{1}{e}[ES \cos(n\Delta t - ES) + EC \sin(n\Delta t - ES)]. \tag{B.66}$$

So the sign of $\sin M$ is the same as the sign of the quantity in the brackets on the right of Eq. B.66. If e is smaller than some quantity, such as 0.1, it is quicker to let $x_0 = n\Delta t$.

Once $x = \Delta E$ has been found, the **f** and **g** functions and their derivatives can be found from Eq. B.41, or

$$\mathbf{f} = \frac{a}{r_0}(\cos \Delta E - 1) + 1,$$

$$\mathbf{g} = \Delta t + \frac{1}{n}(\sin \Delta E - \Delta E),$$

$$\dot{\mathbf{f}} = -\frac{a}{r}\frac{a}{r_0}n \sin \Delta E,$$ (B.67)

$$\dot{\mathbf{g}} = \frac{a}{r}(\cos \Delta E - 1) + 1.$$

To find $\dot{\mathbf{f}}$ and $\dot{\mathbf{g}}$, $r = r(t)$ is needed. This might be found from

$$\frac{r}{a} = 1 - e \cos E$$

$$= 1 - e \cos(\Delta E + E_0)$$

$$= 1 - EC \cos \Delta E + ES \sin \Delta E.$$ (B.68)

But Kepler's equation in the form Eq. B.63 has been solved, and $f'(x)$ is equal to r/a. So this quantity, as well as $C = \cos x$ and $S = \sin x$, can be taken from the subroutine for solving Kepler's equation and used in the construction of Eq. B.67. Finally, $\vec{r}(t)$ and $\vec{v}(t)$ are found from

$$\vec{r}(t) = \mathbf{f}\vec{r}_0 + \mathbf{g}\vec{v}_0, \ \vec{v}(t) = \dot{\mathbf{f}}\vec{r}_0 + \dot{\mathbf{g}}\vec{v}_0.$$ (B.69)

B.9   *Solution in Power Series*

In some programs positions in a Keplerian orbit must be calculated for many times over a short time span. Then the use of the **f** and **g** series is far more efficient than using Kepler's equation, since only polynomials are evaluated. However, more terms may be needed than were derived in Eq. B.44. Solving differential equations numerically using power series can be a useful technique to acquire. This section affords an introduction to it.

Substituting Eq. B.38 into Eq. B.14, and noting that $\vec{r}_0$ and $\vec{v}_0$ are arbitrary vectors, we see that **f** and **g** must satisfy the differential equations

$$\frac{d^2\mathbf{f}}{dt^2} = -\mu\frac{\mathbf{f}}{r^3},$$

$$\frac{d^2\mathbf{g}}{dt^2} = -\mu\frac{\mathbf{g}}{r^3}.$$ (B.70)

Using polar equations in the plane of the orbit,

$$\frac{d^2r}{dt^2} - r\left(\frac{df}{dt}\right)^2 = -\frac{\mu}{r^2},$$

or

$$\frac{d^2r}{dt^2} - \frac{h^2}{r^3} = -\frac{\mu}{r^2}.$$ (B.71)

Let

$$s = \frac{1}{r^3}. \qquad \text{(B.72)}$$

Then

$$\frac{d^2\mathbf{f}}{dt^2} = -\mu\mathbf{f}\, s,$$

$$\frac{d^2\mathbf{g}}{dt^2} = -\mu\mathbf{g}\, s,$$

$$\frac{d^2 r}{dt^2} = +\mu r\, s - h^2 s, \qquad \text{(B.73)}$$

$$3\frac{dr}{dt}s + r\frac{ds}{dt} = 0.$$

The series will be in powers of $(t - t_0)$; we can let $t_0 = 0$ without loss of generality. Let

$$\mathbf{f} = f_0 + f_1 t + f_2 t^2 + \cdots + f_n t^n + \cdots,$$

$$\mathbf{g} = g_0 + g_1 t + g_2 t^2 + \cdots + g_n t^n + \cdots,$$

$$r = r_0 + r_1 t + r_2 t^2 + \cdots + r_n t^n + \cdots, \qquad \text{(B.74)}$$

$$s = s_0 + s_1 t + s_2 t^2 + \cdots + s_n t^n + \cdots,$$

We follow the classic technique of undetermined coefficients. Substitute into the first differential equation

$$2 \cdot 1 f_2 + 3 \cdot 2 f_3 t + 4 \cdot 3 f_4 t^2 \cdots + n(n-1)f_n t^{n-2} + \cdots$$
$$+ \mu(s_0 + s_1 t + s_2 t^2 + \cdots + s_{n-2} t^{n-2} + \cdots)$$
$$\times (f_0 + f_1 t + f_2 t^2 + \cdots + f_{n-2} t^{n-2} + \cdots) = 0.$$

Now pick out the terms factored by t^{n-2}, and equate them:

$$n(n-1)f_n = -\mu(f_0 s_{n-2} + f_1 s_{n-3} + \cdots + f_k s_{n-2-k} + \cdots + f_{n-2} s_0). \quad \text{(B.75)}$$

Similarly,

$$n(n-1)g_n = -\mu(g_0 s_{n-2} + g_1 s_{n-3} + \cdots + g_k s_{n-2-k} + \cdots + g_{n-2} s_0), \quad \text{(B.76)}$$

and

$$n(n-1)r_n = \mu(r_0 s_{n-2} + r_1 s_{n-3} + \cdots + r_k s_{n-2-k} + \cdots + r_{n-2} s_0) - h^2 s_{n-2}. \quad \text{(B.77)}$$

Substituting into the last of the four differential equations,

$$3(r_1 + 2r_2 t + \cdots + (n-1)r_{n-1} t^{n-2} + \cdots)(s_0 + s_1 t + \cdots + s_{n-2} t^{n-2} + \cdots) +$$
$$(r_0 + 2r_1 t + \cdots + r_{n-2} t^{n-2} + \cdots) \times$$
$$(s_1 + 2s_2 t + \cdots + (n-1)s_{n-1} t^{n-2} + \cdots) = 0.$$

Again, equating coefficients of t^{n-2}, we have

$$3(s_0(n-1)r_{n-1} + s_1(n-2)r_{n-2} + \cdots + s_k(n-1-k)r_{n-1-k} + \cdots + s_{n-2}r_1) +$$
$$(r_0(n-1)s_{n-1} + r_1(n-2)s_{n-2} + \cdots + r_k(n-1-k)s_{n-1-k} + \cdots + r_{n-2}s_1) = 0.$$
(B.78)

A computation will begin with values of $\vec{r}$, $\vec{v}$, μ, and $nMax$, or the highest power of t to be computed. From these, find

$$h^2 = (\vec{r} \times \vec{v}) \cdot (\vec{r} \times \vec{v}).$$
(B.79)

Then the series are initialized:

$$\left. \begin{array}{ll} r_0 = \|\vec{r}\|, & r_1 = \vec{r} \cdot \vec{v}/r_0, \\ f_0 = 1, & f_1 = 0, \\ g_0 = 0, & g_1 = 1, \\ s_0 = 1/r_0^3. & \end{array} \right]$$
(B.80)

The principal loop takes n from 2 to $nMax$; each time, finding f_n, g_n, r_n, and s_{n-1}. This is illustrated in the following code. **Hs** $= h^2$.

```
FOR n := 2 TO nMax DO
BEGIN
    f[n] := 0;  g[n] := 0;  r[n] := 0; {Initialize the coefficients.}
    FOR k := 0 TO n−2 DO
    BEGIN
        f[n] := f[n] + f[k]*s[n − 2 − k];
        g[n] := g[n] + g[k]*s[n − 2 − k];
        r[n] := r[n] + r[k]*s[n − 2 − k];
    END; {k loop.}
    f[n] := −Mu*f[n]/(n*(n−1));
    g[n] := −Mu*g[n]/(n*(n−1));
    r[n] := (Hs*s[n − 2] + Mu*f[n])/(n*(n−1));
    FOR k := 0 TO n−2 DO
        s[n−1] := s[n−1] + (n − 1 − k)*(3*s[k]*r[n − 1 − k]
        + r[k]*s[n − 1 − k];
    s[n − 1] := −s[n − 1]/(r[0]*(n − 1));
END; {n loop.}
```

For a given time interval dt, the series should be solved by *back-substitution*; i.e.

```
FSum := f[Nmax];
FOR n := 1 TO Nmax DO
    FSum := dt*FSum + f[Nmax − n];
```

B.10   *The Geometrical Elements of an Orbit*

To interpret the character of a Keplerian orbit, some set of geometrical elements is essential. The elements are used in catalogs, and, more to the point here, will frequently be used as input parameters. A perturbed orbit can also be interpreted by following the *osculating* elements.

The elements fall into two catagories. In the first, the size and shape are given together with information sufficient to locate position in the orbit at any time. These may be

a, the semimajor axis,

e, the eccentricity,

T, the time at pericenter.

For some orbits, such as those of comets, q, the pericenter distance is preferable to a. Also, in place of T, the mean anomaly, M_0, at some specified epoch t_0 may be given; this is usual in the orbits of minor planets.

In the second category are three Eulerian angles specifying the orientation of the orbit in space. These require, first, a *reference plane*. This may be the celestial equator, the ecliptic, the plane of the galaxy, or whatever is appropriate to the context. They also require a given direction in that reference plane; this is the x-axis in Figure B.4. To put the angles into a definite context, let us assume that the reference plane is the ecliptic and that the x-axis points toward the vernal equinox. The *ascending node, N*, of the orbit is the point at which it crosses the ecliptic, going north.

Ω is the *longitude of the ascending node*. It is the angle measured eastward from the vernal equinox to the ascending node.

i is the *inclination*. It is the angle between the ecliptic and the plane of the orbit such that $0 \leq i < 180°$. If $0 \leq i < 90°$, then the orbit is *direct*. If $90° < i < 180°$, then the orbit is *retrograde*.

ω is the *argument of pericenter*. It is the angle measured in the direction of the orbital motion, from the ascending node to the direction of pericenter.

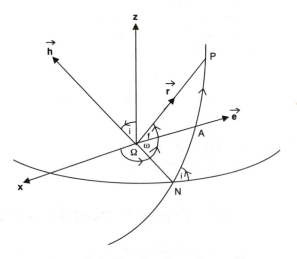

Figure B.4: The elements of an orbit.

Ω and i fix the plane of the orbit in space. ω specifies the location of the orbit in that plane.

For computation, two problems must be addressed. The first is "Given a set of elements, and a time t, how do we find $\vec{r}(t)$ and $\vec{v}(t)$?" The reference system is that shown in Figure B.4, with the z-axis pointing to the north pole of the ecliptic. The recommended procedure (which is far from unique) is first to find components of position and velocity in the orbital reference system at time T, or the given epoch, t_0. These components are transformed to those in the given reference system by three rotations:

1. A rotation through $-\omega$ about the z-axis. The local z-axis, at this stage, is in the orbital reference system and is parallel to $\vec{h}$.

2. A rotation through $-i$ about the x-axis. The local x-axis, at this stage, points along the ascending node.

3. A rotation through $-\Omega$ about the z-axis.

In the programs these rotations are carried out by elementary matrix operations, and should be self-evident. Once components are known in the required coordinate system, the operations of the program for the initial value problem can be used to move to any other required time.

The second problem is "Given components of position and velocity, how do we find a set of elements?" This requires a more careful formulation. Firstly, the three components of $\vec{h}$ in the given reference system can be expressed as

$$\left.\begin{array}{l} h_x = h \sin \Omega \sin i, \\[4pt] -h_y = h \cos \Omega \sin i, \\[4pt] h_z = h \cos i. \end{array}\right] \tag{B.81}$$

Many of the programs use a procedure where the input quantities are proportional to $\sin \theta$ and $\cos \theta$ and the output is θ, where $0 \leq \theta < 2\pi$. This procedure is assumed in the formulation that follows. We are given the time t, $\vec{r}(t)$ and $\vec{v}(t)$.

1. Find a from $a^{-1} = \dfrac{2}{r} - \dfrac{v^2}{\mu}$. This is assumed positive.

2. Find E from

$$e \cos E = 1 - \frac{r}{a}, \quad e \sin E = \frac{\vec{r} \cdot \vec{v}}{\sqrt{a\mu}},$$

and then T from

$$T = t - (E - e \sin E) \sqrt{\frac{a^3}{\mu}}.$$

(You can, of course, add or subtract an integral multiple of the period.)

3. Find the components, (h_x, h_y, h_z), of $\vec{h} = \vec{r} \times \vec{v}$.

4. $\sin \Omega$ is proportional to h_x and $\cos \Omega$ is proportional to $-h_y$. Find Ω.

5. $\sin i$ is proportional to $(h_x^2 + h_y^2)^{1/2}$ and $\cos i$ is proportional to h_z. Find i.

The vector pointing toward pericenter with absolute value e is

$$(e_x, e_y, e_z) = \frac{\vec{v} \times \vec{h}}{\mu} - \hat{r}. \tag{B.82}$$

6. Find (e_x, e_y, e_z) and then $e = (e_x^2 + e_y^2 + e_z^2)^{1/2}$.

Let the unit vector $\hat{n}$ point toward the ascending node, so that $\hat{n} = (\cos \Omega, \sin \Omega, 0)$. Then, since $\hat{n} \cdot \vec{e} = e \cos \omega$,

$$e \cos \omega = e_x \cos \Omega + e_y \sin \Omega. \tag{B.83}$$

Also

$$\hat{n} \times \vec{e} = (e \sin \omega)\vec{h}/h,$$

and

$$\hat{n} \times \vec{e} = (e_z \sin \Omega, -e_z \cos \Omega, e_y \cos \Omega - e_x \sin \omega).$$

Equating the x- or y-components, we find

$$e \sin \omega = \frac{e_z}{\sin i}. \tag{B.84}$$

7. Use the values of $e \cos \omega$ and $e \sin \omega$ to find ω.

Some of these steps are invalid if $i = 0$. If this is the case, then take $\Omega = 0$ (since it is not specifically defined), so that

$$e \cos \omega = e_x, \quad \text{and} \quad e \sin \omega = e_y.$$

B.10.1   Perturbed Keplerian Motion

Usually Keplerian motion must be modified to allow for the effects of perturbing forces. That is, it is necessary to solve the equation

$$\frac{d^2\vec{r}}{dt^2} + \mu \frac{\vec{r}}{r^3} = \vec{F}. \tag{B.85}$$

In most of these programs the equations are solved numerically, using Cartesian coordinates. At every time step the formulas of the preceding section can be used to calculate a set of Keplerian elements. These are called *osculating* elements; they will vary with the time. Following their variation provides geometrical insight into the effects of the perturbing force.

It is also possible to write down explicit equations for the rates of change of these osculating elements. There are various forms of these equations, as can be seen in J. M. A. Danby.[1] Computing with these equations can be more elaborate than using the approach just described; however, a modification of them is used in the module on tidal perturbations in a binary system.

This method of describing perturbations was introduced by Newton, first formulated analytically by Euler and systematically set out by Lagrange. Lagrange uses partial derivatives of the perturbing force $\vec{F}$ with respect to the elements. Gauss produced a form of the equations where $\vec{F}$ is resolved in a coordinate system rotating with the orbital motion. The method is referred to as "variation of arbitrary constants." In fact it is an example of the method of variation of parameters.

The method in these programs for the numerical solution of differential equations is the Runge-Kutta-Fehlberg method with variable stepsize of order five. For details, see E. Fehlberg.[2]

References

1. Danby, J. M. A. *Fundamentals of Celestial Mechanics*, 2nd ed. Richmond, VA: Willmann-Bell, 1988, chap. 11.

2. Fehlberg, E. Low order classical Runge-Kutta formulas with stepsize control. NASA TR 315, 1969.

Index